Psychotherapy for Climate Change explores the complex ways in which climate change is both exacerbating and being fueled by modern society's epidemic of loneliness and social isolation.

Chapters include both research and personal accounts to help the reader deepen their understanding of different perspectives on climate change. The book makes the case for using tools from various types of cognitive behavioral psychotherapy to foster collaboration in addressing the climate crisis.

This book offers exercises that will help readers gain insight into the ways in which emotions and patterns of thinking can impact their capacity for connecting with others about environmental concerns. The book also presents practical tools for strengthening social connections among those who want to contribute to a solution and between environmental advocates and those who deny that the problem exists.

Jennifer L. Keluskar, PhD, is a featured blog writer for *Psychology Today*, a New York State licensed clinical psychologist, and an assistant teaching professor of biobehavioral health at Pennsylvania State University.

"Dr. Keluskar presents excellent insights into the process of coming together in the face of the climate crisis and other forms of existential distress. She weaves her personal journey of self-discovery with a deep dive into research and sheds light on the importance of understanding human diversity in working together towards solutions. I highly endorse it."

Linda Goldman, author of *Climate Change and Youth: Turning Grief and Anxiety into Activism*

"In this timely and insightful book, Jennifer Keluskar offers a clear, reasoned, and practical guide to the practice of climate-informed psychotherapy that skillfully weaves a compelling narrative about ways to build resilience. This book is an essential resource for anyone working at the intersection of mental health and our changing planet."

James McKowen, PhD, clinical psychologist and co-chair of the Massachusetts General Hospital Climate Mental Health Initiative

"In *Psychotherapy for Climate Change*, Jennifer Keluskar applies her knowledge of dialectical behavior therapy to help people understand and overcome perfectionism and needless suffering about climate change and move toward healthy feelings, values, and actions. This is a great resource for cognitive behavioral therapists and their clients."

Thomas Doherty, PsyD, climate psychology researcher and author of *Surviving Climate Anxiety*

"Jennifer Keluskar has written a wonderful book on cognitive behavioral therapy approaches to help address the realities of the climate crisis. This is a great read for clinicians and non-clinicians wanting to learn more about psychological responses to climate change."

Jesse Homan, PhD, LPC, co-chair of the Climate Psychology Alliance Expressive Arts Committee

PSYCHOTHERAPY FOR CLIMATE CHANGE

A PRACTICAL GUIDE TO TRANSCENDING DIFFERENCES

Jennifer L. Keluskar

Routledge
Taylor & Francis Group
NEW YORK AND LONDON

Designed cover image: Cover art courtesy of Pam Longobardi (2013). *Economies of Scale* [microplastics, plastic, hydrocarbons, steel]. Palmer Museum of Art, State College, Pennsylvania, USA.

First published 2027
by Routledge
605 Third Avenue, New York, NY 10158

and by Routledge
4 Park Square, Milton Park, Abingdon, Oxon, OX14 4RN

Routledge is an imprint of the Taylor & Francis Group, an informa business

ISBN: 9781032776668 (hbk)
ISBN: 9781032776651 (pbk)
ISBN: 9781003489665 (ebk)

DOI: 10.4324/9781003489665

Typeset in New Baskerville
by KnowledgeWorks Global Ltd.

To Skylar, Colby, and Raja

Contents

List of Figures and Table

FIGURES

TABLE

Acknowledgments

This book would never have been possible without the support of my colleagues, friends, and family.

I am grateful to my colleagues for their generosity and enthusiasm for my vision. Linda Goldman encouraged me to pursue this project and offered insights that enhanced several chapters. Jesse Homan shared thoughtful and thorough feedback for Chapter 7 on Dialectical Behavior Therapy. Kelsey Hudson provided her expertise and suggestions for Chapter 8 on Acceptance and Commitment Therapy. Leo Goldsmith offered extensive guidance on sections discussing LGBTQIA+ identities and environmental activism.

James McKowen, Mor Keshet, Doug Wrenn, and Lauren Krueger allowed me to interview them, and I am incredibly appreciative of their wisdom, help, and time.

Sami Aaron reviewed information I gleaned from past conversations and allowed me to discuss her vision for caring for climate activists within these pages.

Kristina Pikunas provided mentorship and encouragement throughout this project.

Climate scientist and activist Peter Kalmus and philosopher Bill Irvine, both highly acclaimed in their professions, took the time to review my references to correspondence with them.

My thanks to Anna Moore and the other members of the Routledge editorial and production team.

Of course, this project would also never have been possible without my family, so I want to acknowledge Skylar and Colby for their boundless curiosity and forgiveness, and Raja for fixing problems and not fixing them when I just needed someone to listen.

Finally, I want to thank my parents. My dad taught me everything I know about the writing process, and my mom never says no to being a sounding board to vent to. Both gave me my stubborn persistence, which allowed me to complete this and other meaningful endeavors.

About the Cover Art

Longobardi, P. (2013). *Economies of Scale* [microplastics, plastic, hydrocarbons, steel]. Palmer Museum of Art, State College, Pennsylvania, USA.

Pam Longobardi's sculpture, *Economies of Scale,* consists of ocean plastic removed from several different countries. The sculpture makes a statement about the urgent need to consider the costs of polluting the earth's oceans. The term *Economies of Scale* comes from microeconomics and refers to the idea that increased output of a product confers a cost advantage.

The sculpture appealed to me due to its environmental theme as well as the idea that, while linking together ocean waste symbolizes the dire nature of our environmental problems, the solution explored in this book lies in linking people together in collective efforts.

Introduction

This book will provide suggestions for people who are openly struggling with eco-distress, but it will also speak to the general confusion and dread felt by the larger body of eco-conscious individuals.

In a way, my motivation for writing about this subject can be summarized by an article about paper towels. The article was featured in *Wirecutter*, the product review section of *The New York Times*. Author Annmarie Conte (2024) entertained a question from a reader, who asked how they could get their boyfriend to stop using so many paper towels. Conte's response includes a range of behavior modification suggestions, such as keeping the paper towels out of sight, and keeping cloth towels organized, clean, and more accessible. Although it was refreshing to read about tips for environmental sustainability, I was intrigued by some of the readers' comments.

According to one reader, "Paper towels are better for the environment than wasting all the water and energy on washing fabric towels. Also, used paper towels are better for compost."

Although the environmentalists reading this might roll their eyes, this is similar to a conversation I had with a friend in graduate school, when we debated the merits of using non-disposable thermoses for coffee given that we had to wash them. It's obvious now that this is more environmentally sustainable than disposable cups, but it's important for eco-conscious people to realize that these questions are not necessarily obvious to people who haven't yet done the research on sustainable living. Even the highly educated can be misinformed about energy and resource use.

DOI: 10.4324/9781003489665-1

Comments in response to the Wirecutter (Conte, 2024) piece also revealed germophobic sentiments so close to my own struggles:

> ... If you are going to re-use towels, make sure you understand what 'cross contamination' is and be careful about it .... One trip to the ER costs a lot more (money, resources) than a few paper towels when you cook raw chicken and beef.

What fascinates me about these comments is that they reveal a spectrum of responses we have about environmental sustainability. The responses go well beyond the dichotomy of activist versus denier. We see the reactions that come from a place of disinformation (i.e. that washing a re-usable is worse for the environment), and the impatience it strikes in those who know better. It shows the conflicts that come up even in those who are knowledgeable *and* care, but who struggle with the real possibility that many of our sustainable efforts are sabotaged by unpredictable factors, like contagion, illness, and broken appliances that need to be tossed.

This book aims to address these divisions with the hope that individual mental health practices can contribute to the cultural change we need to begin to reverse our climate catastrophe.

There are times I feel optimistic about the potential for unity among climate advocates, such as on September 17, 2023, when I participated in the March Against Fossil Fuels in New York City. It was my first protest since college, when I was part of the Student Global Aids Campaign. Despite my initial anxiety about thinking I was too old to be a protestor, to the contrary, I saw people of all ages and backgrounds at the event. Young adults from the Sunshine Movement danced and sang on top of a school bus. Middle aged adults held posters on the streets, while their children played or were carried on their backs. Elders marched with their young adult children. Peter Kalmus, a well-known climate scientist and activist, spoke at the event, as did American politician and activist Alexandra Ocasio-Cortez. Within the crowd though, scientists and laypeople blended together seamlessly.

While a one-day show of solidarity can easily unite people from all walks of life, the question of how we can become a collaborative, unified front in our daily lives is more complicated. This concept is the essence of this book. An example is the story behind *Energetic*, an ingenious board game developed to help professionals understand New York City's energy challenges (Krueger, 2023). In this game, players choose among one of four roles: Politician, engineer, entrepreneur, and activist, which reminds me of the disenfranchisement I have felt in my own climate advocacy efforts, as I do not fall into any of these categories. In the real world, there needs to be a role for everyday consumers, and that is an important part of what this book is about.

This book aims to help people of diverse backgrounds feel like they have a place at the table of climate advocacy. It is written in a way that people without a background in clinical psychology can understand and benefit from. Similar to a self-help book, it provides examples from my clinical work and stories from my personal life that highlight skills that I further break down into their component parts, thereby also potentially being of interest to mental health practitioners. It shows how climate change comes up in many life stories as the backdrop to more immediate concerns, but one that, in being addressed, can open the door for better mental health.

WHAT TO EXPECT FROM THIS BOOK

The first 5 chapters this book explore the diverse ways people have responded to the climate crisis, as well as the various factors that contribute to these responses, including personal identities, world views and ideologies, belief systems, and spirituality. Throughout, I highlight the challenges of different perspectives in these areas. However, sprinkled throughout the first 5 chapters of this text, I introduce strategies that will be explored further in later chapters.

RATIONALE FOR DEVELOPING COGNITIVE BEHAVIORAL INTERVENTIONS FOR ECO-DISTRESS

A 2024 *New York Times* piece (Caron, 2024), entitled "Global Issues are Taking a Toll on Youth's Mental Health" highlights the need to look beyond the individual in addressing psychological distress. The piece calls for overhauling how we approach mental health, citing a policy brief on a *Lancet* study examining trends in mental health of teens and young adults (2024). While financial problems, including the cost of housing and job insecurity, are among the stressors exacerbating mental health in youth, megatrends such as climate change are contributing to a sense of an uncertain future as well. The brief report notes: "It is no exaggeration to say young people and their mental health act as the early warning system for contemporary society; they are manifesting the warning signs and symptoms of a society and world that is in serious trouble" (p. 2).

How does a psychotherapist equip their clients to successfully navigate the challenges of a sick society within the paradigm of individual psychotherapy? This is a complicated but important question.

The idea that institutions influencing government policy filter down to affect psychological and emotional well-being is consistent with Urie Bronfenbrenner's ecological systems theory of child development (Bronfenbrenner, 1979), according to which human development is affected by different levels of influence, including the *microsystem, mesosystem, exosystem, macrosystem,* and *chronosystem.*

Urban planning that prioritizes access to green spaces (Nguyen et al., 2021) can be applied to Bronfenbrenner's (1979) model. Children playing together in their backyard or neighborhood park constitutes the microsystem because it involves direct interaction with their immediate environment. The mesosystem, which represents the interaction of microsystems, would be represented by parents and teachers collaborating to provide more outdoor learning experiences. Aspects of urban planning that allow for enjoying the outdoors, such as bike paths, require the approval of local governing bodies, which are included in the exosystem, which encompasses the decisions that are made in a setting with which the child doesn't have direct interaction. Social norms and broader cultural beliefs for outdoor recreation are representative of the macrosystem. The chronosystem represents changes over time, including events that affect a child's development. In the context of the climate crisis, this can be seen in younger generations expressing more anxiety about how climate change will affect their quality of life. These examples show how, according to Bronfenbrenner's (1979) model, different layers of the ecological system interact with each other. For instance, parent-child interactions (described later in this book) can interact with other layers to affect development.

Bike paths run through my town, and this has motivated me to bike to work. Doing so has allowed me to exercise, get fresh air, and enjoy the beauty of green space, all of which has been great for my mental health. In contrast, toxins such as particulate matter in the atmosphere, unchecked use of pesticides, lack of green spaces, and rising temperatures, all of which are exacerbated by fossil fuel combustion, have been linked to adverse health outcomes (e.g. Calderón-Garciduneñas et al., 2021). As discussed later in this book, some individuals have more exposure to these toxins based on where they live and other factors, but ultimately the cleanliness of the environment will affect everyone's health.

When I started talking to people about the idea for writing a book about using psychotherapy to address our response to the climate crisis, it was met with resistance by those who balked at the idea of "putting a band aid" on the issue or blaming the individual for the inadequacy of institutions. Climate change is a political problem, not an individual's psychological one, they argued.

These sentiments are valid, but the purpose of this book is to join a growing movement of mental health providers and researchers who are striving to elevate psychotherapy and transcend the individual level to address societal woes, including those such as racism, sexism, heterosexism, and cisgenderism, and other social inequalities that intersect with the climate crisis.

When we broaden our understanding of wellness from the focus of the individual to also include a healthier society, we help the individuals who are part of it. As stated by LGBTQIA+ activist and scholar Leo Goldsmith, whom I interview later in the book, therapy *is* activism because it is helping others (personal communication, February 25, 2025). As a profession, we can take this a step further by honing tools to help clients become part of the solution to social injustice issues and model a spirit of advocacy, which helps clients feel validated and less isolated.

A risk of separating broader societal problems from mental health is ignoring the struggles faced by people in disadvantaged situations, or assuming they are too preoccupied with life-threatening challenges to be concerned with emotional ones. A case in point is the unmet need for treatment of post-traumatic stress disorder (PTSD) among Palestinian refugees (Asi, 2024). The refugees of Gaza are facing imminent death from bombings, starvation, and disease. While physical needs should be prioritized, their mental health struggles also should not be forgotten.

While climate distress is existential at its core, that does not mean it is any less real. The effects of climate change are already exacerbating natural disasters, like drought, flooding, and air pollution, that are contributing to suffering. Our unremitting consumption of fossil fuels is putting us on the fast track to early death, and it is a detriment to health and livelihood over which we have limited control. One could choose to stop smoking, but they can't control the particulate matter they are breathing in from surrounding traffic.

If a person were to tell you that they were scared of death, we would not say that they have a mental health disorder. However, if their fear of death fueled excessive avoidance, such as avoiding leaving their house, even to run errands, out of fear, then we would be more concerned. It makes sense that cognitive behavior therapy (CBT) for eco-distress need not invalidate the real danger of climate change and other environmental threats.

CBT, like other psychotherapeutic orientations, can validate the client's existential angst, help the client understand what is blocking them from pro-environmental behavior and participating in collective action, and teach techniques for behavior change when eco-distress is so severe as to be limiting their functioning. For example, research (Ogunbode et al., 2022) shows that while climate anxiety is associated with individual pro-environmental behavior, such as recycling, and environmental activism, it is also related to less mental well-being. This is concerning given that we need those who are behind such a critical cause to be in good health, both physically and psychologically.

THE CHALLENGES OF ASSESSMENT

I've hypothesized reasons clients might feel reluctant to present with eco-distress. It is a problem that has been invalidated, and perceived as insolvable. I've also heard concerns among colleagues that assessing eco-distress might inadvertently lead to clients feeling judged for not being concerned enough. This leads us to question whether eco-distress should be assessed at intake to begin with. While the clinician can subtly prompt the client to express eco-concerns by labeling themselves as a climate-aware therapist, such a "wait and see" notion might be as naïve as expecting clients to tell you everything that might be relevant to their care without prompting.

Measures such as the Climate Change Anxiety Scale (Clayton & Karazia, 2020) can be included in an intake packet and follow up questions to the client's response can help paint a clearer picture of the role eco-distress is playing in clinical presentation. Alternatively, the clinician can assess the client's environmental identity, including connection to nature (Doherty et al., 2024). Doing so might not only accomplish picking up on eco-distress but also inform nature-based interventions, which have been shown to be beneficial for social and emotional well-being.

THE TARGETS OF TREATMENT

In *Psychiatric Perspectives on Youth Climate Distress Using the Biopsychosocial-environmental Knowledge Base to Understand and Assess for Clinical Level Symptoms,* Elizabeth Haase (2024) outlines clinical questions for eco-distress in young people, including whether research highlighting the prevalence of climate distress in youth is what we would deem clinically significant, or significantly above the norm; whether and how pre-existing psychiatric diagnoses are related to climate distress; how the outcome of treatment should be measured given that our goal is not to completely mitigate climate distress (as this distress is valid and can motivate the desire to advocate for change); and whether the common recommendation of taking meaningful action is helpful in bringing about desired outcomes.

Recent research is starting to shed light on these essential questions, although much still needs to be done to gather enough information to create a roadmap for psychotherapists in treating the eco-distressed. A longitudinal study by Veijonaho et al. (2025) showed that while adolescents with more climate distress engaged in more pro-environmental behavior contemporaneously, higher distress was associated with less than expected pro-environmental behavior one year later. In turn, engaging in higher than expected levels of pro-environmental behavior at one time-point was associated with higher distress one year later.

This study (Veijonaho et al., 2025) also examined how different coping styles related to distress and behavior affected pro environmental behavior and subsequent distress. Two of the styles that were explored were problem-focused coping, which entails taking action, such as via pro-environmental behaviors, and meaning-focused coping, or engendering a sense of meaning to fuel motivation, believing in collective action, and nurturing hope in working together toward a solution. The study showed that problem-focused coping alone led to less pro-environmental behavior and more distress in the longer term, but when it was coupled with meaning-focused coping, climate distress remained stable in the longer term. The authors suggest that meaning-focused coping serves to build a sense of agency, which buffers the individual from negativity.

Another reason there is still much to be discovered in psychotherapy for eco-distress is the question of *what* we are treating. Given that eco-distress is not a disorder that is part of the Diagnostic and Statistical Manual of Mental Disorders – Fifth Edition (DSM-V), the targets of treatment will vary depending on which aspect of distress about environmental devastation we're addressing considering that environmental devastation can take various forms, from climate change and its consequences, to concern about toxins, to grief over the destruction of wildlife. Furthermore, these targets need to be considered within the context of other psychological factors including pre-existing mental health disorders, personality traits, values, and the history of life events that have shaped the individual.

What would the typical client with eco-distress look like? There isn't a clearcut answer to this question. In some cases, it might be an activist who is struggling with the frustrations of their labor not bringing about change in the proportion of exerted effort. This might be a case of burnout that is exacerbated by invalidating social institutions and stereotyping of activists as angry lawbreakers. On the other hand, an activist who is frustrated about lack of cooperation from their followers might benefit from exploring different interpersonal styles to better gain commitment from others.

Alternatively, the client might not be actively engaged in climate activism but might present with a previous mental health diagnosis into which eco distress has been folded, generally via a psychological mechanism such as perfectionism. An example might be someone with obsessive-compulsive disorder (OCD), a condition characterized by intrusive, unwanted thoughts that are often relieved only through compulsive behaviors, such as repetitive checking, and rituals such as repeating a specific phrase (DSM-V-TR, American Psychiatric Association, 2022).

An individual with OCD might find their checking compulsion gets ramped up after they read an article about how the destruction of ecosystems will lead to scarcity of resources. This might manifest in compulsively checking that appliances and faucets are turned off in the home. The individual with Generalized Anxiety Disorder (GAD), characterized by excessive worries about common concerns such as getting sick or academic achievement (DSM-V-TR, American Psychiatric Association, 2022), might include climate change in their entourage of uncontrollable worries. An individual with a history of disordered eating might rationalize restriction of less sustainable foods such as meat and dairy.

We can also see this going in the other direction where a person engaged in climate work hits a roadblock emotionally, such as frustration over the government's inaction, and pre-existing characteristics render it more challenging to stay resilient. These factors could include one's way of thinking or behavioral repertoire, which might interfere with being able to cope when things don't go as planned. For example, while a setback such as not being able to institute a carbon tax is understandably upsetting, people vary in their degree of resilience in response to such challenges. Some might become so distressed as to lose the motivation, energy, and interpersonal skills needed to rise to these challenges. They might be reluctant to take breaks or ask for help, both of which are signs of resilience, not of weakness. And yet many high achievers have trouble embracing them, ironically rendering them less resilient.

In either case, I'm not implying that environmentalism is bad for the individual. And mitigating psychological symptoms, in turn, will strengthen their efforts to advocate for societal change. Greta Thunberg serves as an example of the transformative power that spearheading a cause can have on mental health (BBC, 2024).

We approach these cases not by invalidating people's concerns about climate change, but doing the opposite: Bringing them to light. In order to do this, we need to understand the individual before us, and how they relate to today's environmental problems.

Chapters 1–5 of this book dive into the layers that influence who people are and what makes them tick as the essential foundation for Chapters 6–10, which will dive into specific cognitive-behavioral skills.

Additionally, we need to consider when climate action might be prescriptive for eco-distress. Actions that make a difference can be therapeutic, and pro-environmental behavior and collective actions (e.g. joining an environmental advocacy group) are often cited as the goal for treatment among the eco-distressed (Schwartz et al., 2022). However, for many, this goal is not an endpoint, but rather a beginning, as Veijonaho et al.'s (2025) longitudinal study suggests.

This book is also written for cases in which people do not come to therapy with the intention of treating eco-distress but find it coming up in the course of their sessions, perhaps unexpectedly. As a clinician, I've felt thrown off guard at such moments, and before I started focusing more on existential concerns in my clients, I made mistakes, such as assuring one client that Donald Trump wasn't going to win the 2016 presidential election when they expressed concern about this prospect on election day.

Our instinct is to reassure and soothe when people share their fears of global events, but doing so is not only unhelpful but risks invalidating clients and sending the message, however unintentional, that this is not kind of stuff they are supposed to bring up in a therapy session.

Clients might also feel uncomfortable talking about climate change because unfortunately, it has become tied into politics. Upon seeking further education about the mental health effects of climate change, clinicians can advertise themselves as "climate aware," which signals that they will be validating of climate concerns and that they themselves have environmental concerns.

Let's consider another scenario in which the client has an interest in environmentalism but denies eco-distress at intake. They are coming to you because they are lacking meaning and purpose in their life. Could encouraging pro-environmental action help this individual get out of a rut?

Melissa Bernstein, co-founder of Melissa & Doug toys, had spent much of her life struggling with anxiety and depression, including existential concerns (McCarthy, 2021). As described in a *Washington Post* piece, Bernstein was gifted as a child but felt that the dark thoughts she sometimes had couldn't be shared with anyone. She responded to her depressive tendencies by making sure she always stayed busy. She showed a perfectionistic cognitive style, such as worrying that her children wouldn't stand out as the brightest and best athletes.

In 2017, Bernstein read about existentialism and felt it related to how she had felt much of her life (McCarthy, 2021). She learned that people who struggle with existential depression tend to be both creative and highly sensitive. This aligns with my clinical experience working with gifted children, who felt troubled by topics such as the limited time we have on earth, and being able to make a big enough difference in the world in their lifetime.

Bernstein extended her work to the mental health realm after she wrote a memoir called *LifeLines: An Inspirational Journey from Profound Darkness to Radiant Light* (2021). For a period of time, she offered free mental health resources to the community.

Even if environmentalism isn't your calling, it can be linked to existential dread in acknowledging that such angst is being fueled by a productivity- and status-obsessed society. One solution to climate change is collectively slowing down. The goals of addressing the climate crisis and psychotherapy share an essential core of restoring balance. Going to bed earlier means saving electricity and getting adequate rest, which makes it less likely that we will get distracted and leave the fridge door open. When we are at peace on the inside, we don't need to consume as much (entertainment, substances, etc.).

At the same time, authentically integrating the goals of climate psychology and mental health also demands a paradigm shift. The self-help world is inundated with products that claim to be able to fix your woes, and mental health tools have unfortunately become entwined with commercialism. At the same time, as environmental advocates, it is essential to not deprive ourselves intentionally. Doing so is for one, unsustainable, and also goes against the more powerful cause of alleviating and preventing human suffering that is and will continue to be linked to climate change and other environmental threats.

Even when we attempt to remove the capitalist aspects of mental health, there's still the tempting tie between consumption and contentment. Taking hot showers, watching comedies, drinking coffee, and eating chocolate all fall toward the top of my list of go-to coping tools, but they are all forms of consumption as well. I would argue that doing all the above, *mindfully*, or with full awareness of what your senses are experiencing in the moment, and appreciation for all of the products that are ultimately tied to the earth, will allow us to live more sustainably without feeling deprived.

Being creative will also achieve the end goal of sustainability, both spiritually and environmentally. One of my favorite hobbies is figuring out how to use leftovers in a new recipe, an outlet for creativity as well as an effort to avoid food waste.

Psychotherapy for climate change also means addressing perfectionistic thoughts that one must never waste or pollute, or in other words, never make an environmental mistake. This is an impossible expectation, and one that traditional CBT could address by highlighting unhelpful cognitive styles like all or nothing thinking and catastrophizing without invalidating the clients' aims of sustainability and climate advocacy.

Another type of CBT is Radically Open Dialectical Behavior Therapy (RO DBT), which is an evidence-based treatment that specifically addresses maladaptive perfectionism (Lynch, 2018). RO DBT does not aim to change one's personality fundamentally, but rather to soften it sufficiently to allow the client to feel more connected with others. Their position, which is reflected in others' research on success-driven people, is that at a certain point, striving comes at the cost of human connection.

The climate crisis is a metaphor for the life of a perfectionist. Perfectionism is not sustainable, yet knowing that achievement, correctness, and all the other attributes perfectionists take to an unhealthy extreme are at their core, prosocial, or valued by society, perfectionists might not know how else to live.

One aim of RO DBT is to help people relate with others more flexibly, which is facilitated by thinking more flexibly (Lynch, 2018). For instance, how do I reconcile writing a book about climate change while eating a bag of candy, the individually packaged wrappers piling on top of my nightside table? Resolving contradictions like this by giving up on the book would constitute what RO DBT practitioners refer to as *fatalistic* mind (Lynch, 2018) or throwing in the towel in the face of challenges. It is what RO DBT practitioners refer to as the "escape artist" of the mind.

Fatalistic mind is at play when I feel hopeless about doing the "right" thing environmentally. For example, I'm riding a bicycle to work, but because the daily temperature has risen to the 80s recently, I'm also doing a lot more laundry because of having heaps of sweat-drenched clothing. It feels like an endless, exhausting cycle.

Fixed mind is another form of inflexibility in which the person insists that only one perspective is right (Lynch, 2018). It can be tempting for seasoned environmental activists to fall into this state of mind when talking about climate change. After all, science is as clear as an unpolluted sky. Except being "right" is not going to mobilize the extent of cultural shifts and collective action needed. There are examples in public health advocacy, such as efforts to curb female genital mutilation (FGM) in Africa, where cultural shifts happened more readily in tribes where there was intermarriage with other tribes that didn't practice FGM (Mkuwa et al., 2023). The important lesson here is that diversity of thought, culture, etc. within a community can facilitate flexibility and therefore make necessary changes for improving public health.

Importantly, this book is meant to help mental health clinicians identify the subtle ways that climate change is coming up in clinical work, use it as a tool to help clients increase self-efficacy, and imbue clients with a sense of having greater meaning and purpose in their endeavors.

CBT CASE CONCEPTUALIZATION AND CLIMATE DISTRESS

A cognitive-behavioral case conceptualization is comprised of the following components (Persons, 2008):

Origins: These include any events that affected the individual's development during their childhood years (e.g. trauma, parental modeling, etc.) as well as biological/genetic predispositions (e.g. temperament).

Mechanisms: These include, personality traits and developmental/cognitive or processing differences, as well as beliefs the person has formed about themselves, others, and the world in general. Mechanisms are the main target of treatment in CBT. For instance, avoidance, perfectionism, and compulsive striving are all mechanisms, as are beliefs you must be liked by everyone in order to be worthy, or making a mistake means you are a failure.

Precipitants: These are events that trigger the mechanisms (shaped by origins), all of which lead to the onset and/or maintenance of maladaptive behaviors. An example would be coming home and realizing that you left the fridge door open all day being a precipitant for perfectionistic thoughts that this mistake is evidence of being incompetent.

Problems: While mechanisms are the target of treatment, problems are the reason people come in for psychological treatment. They include mental health symptoms, such as depression, anxiety, etc. and areas of functioning that are affected by them, such as relationship problems, work problems, sleep and/or eating difficulties, etc. They also include physiological arousal and emotions (e.g. guilt, fear, rage) that often accompany maladaptive reactions.

Elizabeth Marks and Kelsey Hudson (2024) present a cognitive behavioral model for understanding what they term impairing eco-distress, such as not being able to focus on daily activities and relationships after hearing about droughts and wildfires. Their conceptualization frames the goal of treatment as constructive eco-distress. It integrates skills from Acceptance and Commitment Therapy (ACT) such as reworking one's relationship with uncomfortable emotions about environmental threats by seeing them as a signal that one cares about the environment.

Note that the premise of the model is to help individuals respond to valid concern about the climate crisis with proactive responses, such as keeping up with both positive and negative news, using self-talk to emphasize that others are also working to improve the situation, engaging in pro environmental behaviors and taking breaks when needed. Their conceptualization portrays CBT as a model of sustainability, as opposed to the band aid it is often criticized as representing.

While the first five chapters of the book dive into various aspects of identity that influence how we respond to climate change and other aspects of environmental devastation, Chapters 6–10 discuss suggestions for applying evidence-based therapies, specifically RO DBT, Dialectical Behavior Therapy (DBT), ACT, and traditional CBT.

We do not yet have studies that explore using these specific modalities for eco-distress. Along these lines, my suggestions don't come from any treatment manual specific to using that treatment for eco-distress. My hope is that they will represent logical and helpful extensions of evidence-based practice, and that researchers might consider these insights in future clinical studies.

CBT is not a band aid for climate distress. It is the means to the inner wisdom that will elevate us toward connection and collective commitment to creating a more sustainable and compassionate world.

REFERENCES

American Psychiatric Association. (2022). *Diagnostic and statistical manual of mental disorders* (5th ed., Text rev.). American Psychiatric Association Publishing.

Asi, Y. M. (2024, February 22). The trauma experienced in Gaza is beyond PTSD. *The New York Times*. https://www.nytimes.com/2024/02/22/opinion/gaza-palestinians-mental-health.html

Bernstein, M. (2021). *LifeLines: An inspirational journey from profound darkness to radiant light.* LLC.

British Broadcast Company (BBC). (2024, May 9). Greta Thunberg: Who is the climate activist and what has she achieved? *BBC.* https://www.bbc.com/news/world-europe-49918719

Bronfenbrenner, U. (1979). *The ecology of human development: Experiments by nature and design.* Harvard University Press.

Calderón-Garciduneñas, L., Stommel, E. W., Rajkumar, R. P., Mukherjee, P. S., & Ayala, A. (2021). Particulate air pollution and risk of neuropsychiatric outcomes. What we breathe, swallow, and put on our skin matters. *International Journal of Environmental Research Public Health, 18*(21), 11568. https://doi.org/10.3390/ijerph182111568

Caron, C. (2024, August 13). Global issues are taking a major toll on young people's mental health. *The New York Times.* https://www.nytimes.com/2024/08/13/well/mind/mental-health-young-adults-trends.html

Clayton, S., & Karazia, B. (2020). Development and validation of a measure of climate change anxiety. *Journal of Environmental Psychology, 69.* https://doi.org/10.1177/0269881115573542

Conte, A. (2024, February 28). Ask Wirecutter: How do I curb my boyfriend's paper towel addiction? *The New York Times.* https://www.nytimes.com/wirecutter/reviews/ask-wirecutter-paper-towels/?searchResultPosition=1

Doherty, T., Artman, S., Homan, J., Keluskar, J., & White, K. E. (2024). Environmental identity-based therapies for climate distress: Applying cognitive behavioral approaches. *The Cognitive Behaviour Therapist, 17,* e29. https://doi.org/10.1017/S1754470X24000278

Gillis, J. (2014, November 2). U.N. panel issues its starkest warning yet on global warming. *Europe. The New York Times.* https://www.nytimes.com/2014/11/03/world/europe/global-warming-un-intergovernmental-panel-on-climate-change.html

Gillis, J. (2015, January 16). 2014 breaks heat record, challenging global warming skeptics. *Environment. The New York Times.* https://www.nytimes.com/2014/11/03/world/europe/global-warming-un-intergovernmental-panel-on-climate-change.html#:~:text=Panel%20Issues%20Its%20Starkest%20Warning%20Yet%20on%20Global%20Warming,-Share%20full%20article&text=COPENHAGEN%20%E2%80%94%20The%20gathering%20risks%20of,major%20new%20United%20Nations%20report

Haase, E. (2024). Psychiatric perspectives on youth climate distress: Using the biopsychosocial-environmental knowledge base to understand and assess for clinical level symptoms. In K. Hudson & E. Haase (Eds.). (2024). *Climate change & youth mental health: Multidisciplinary perspectives.* Cambridge University Press.

Hayes, S. C., & Strosahl, K. D. (Eds.). (2005). *A practical guide to acceptance and commitment therapy.* Springer Science + Business Media. https://psycnet.apa.org/record/2005-02327-000

Janoski, T. P., Broccoli, A. J., Kapnick, S. B., & Johnson, N. C. (2018). Effects of climate change on wind-driven heavy snowfall events over eastern north America. *Journal of Climate, 31*(22), 9037–9054. https://doi.org/10.1175/JCLI-D-17-0756.1

Krueger, A. (2023, April 21). Roll the dice, save Gotham from climate catastrophe! *The New York Times.* https://www.nytimes.com/2023/04/21/nyregion/earth-day-energetic-game.html

Linehan, M. (1993). *Cognitive-behavioral treatment of borderline personality disorder.* Guilford Press.

Lynch, T. (2018). *Radically open dialectical behavior therapy: Theory and practice for treating disorders of overcontrol.* Context Press.

Marks, E., & Hudson, K. (2024). Cognitive behavioral principles for conceptualizing young people's eco-emotions and eco-distress. In K. Hudson & E. Haase (Eds.), *Climate change & youth mental health* (pp. 148–167). Cambridge University Press.

McCarthy, E. (2021, April 27). She brought kids joy with her 'magical toys. And hid her own misery until it became unbearable. *The Washington Post.* https://www.washingtonpost.com/lifestyle/style/melissa-and-doug-toys-founder-depression/2021/04/26/8d4af4be-7ba7-11eb-85cd-9b7fa90c8873_story.html

Mkuwa, S., Sempeho, J., Kimbute, O., Mushy, S. E., Ndjovu, Mfaume, J., & Ngalesoni, F. (2023). The role of communities and leadership in ending female genital mutilation in Tanzania: An exploratory cross-sectional qualitative study in Tanga. *BMC Public Health, 23,* 163(2023). https://doi.org/10.1186/s12889-023-15086-z

Nguyen, P., Astell-Burt, T., Rahimi-Ardabili, H., & Feng, X. (2021). Green space quality and health: A systematic review. *Environmental Research and Public Health.* https://doi.org/10.3390/ijerph182111028

Ogunbode, C. A., Doran, R., Hanss, D., Ojala, M., Salmela-Aro, K., van den Broek, K. L., Bhullar, N., Aquino, S. D., Marot, T., Schermer, J. A., Wlodarczyk, A., Lu, S., Jiang, F., Maran, D. A., Yadav, R., Ardi, R., Chegeni, R., Ghanbarian, E., Zand, S., …Karasu, M. (2022). Climate anxiety, wellbeing and pro-environmental action: correlates of negative emotional responses to climate change in 32 countries. *Journal of Environmental Psychology, 84,* 101887.

Persons, J. B. (2008). *The case formulation approach to cognitive-behavior therapy.* Guilford University Press.

Rauh, V. A., & Margois, A. (2017). Research review: Environmental exposures, neurodevelopment and child mental health – New paradigms for the study of brain and behavioral effects. *Journal of Child Psychology and Psychiatry, 57*(7), 775–793. https://doi.org/10.1111/jcpp.12537

Schwartz, S. E. O., Benoit, L., Clayton, S., Parnes, M. F., Swenson, L., & Lowe, S. R. (2022). Climate change anxiety and mental health: Environmental activism as buffer. *Current Psychology, 28,* 1–14. https://doi.org/10.1007/s12144-022-02735-6

Veijonaho, S., Hietajärvi, L., Ojala, M., & Salmela-Aro, K. (2025). From distress to action? – A three-wave longitudinal study of climate change distress, pro-environmental behavior, and coping strategies among Finnish adolescents. *Journal of Environmental Psychology, 104,* 1–11. https://doi.org/10.1016/j.jenvp.2025.102676

Chapter 1

The Seven Responses to Climate Change

THE FIRST STEP TO CHANGING THE WORLD IS UNDERSTANDING THE DIVERSITY OF HUMANKIND

In a May 2023 episode of Julia Louis Dreyfus' podcast *Wiser Than Me*, Dreyfus interviews Gina McCarthy, whom Dreyfus refers to as the "MacGyver of Climate" (6:48). McCarthy was the head of the Environmental Protection Agency (EPA) during the Obama administration, the first National Climate Advisor under the Biden administration, and president of the Natural Resources Defense Council (Dreyfus, 2023; ginamccarthy.com; NRDC, 2025). Her success in propelling environmental wins, such as spearheading initiatives to protect natural resources and reduce greenhouse gases, was fueled by her passion for collaboration. McCarthy's success is largely defined by her ability to rise above the political polarization that has become all too common in modern America.

Climate change and other environmental threats trigger intense emotions whether you deny their existence or advocate for solutions. Either way, you are presented with an enormous demand to change everything you know about the way you live, a change so great that you won't be able to do it alone. While climate deniers push back against the demand, advocates beseech others to get on board with the broad societal changes that are needed for enough to be done. Emotional responses to how to go about change make it challenging for people to say or do anything that might suggest they are budging from what they "know is right," which in turn fuels frustration with those whose beliefs are dialectically opposed to theirs.

One way in which climate change affects mental health is in the tension surrounding the very topic. Controversy breeds isolation and

DOI: 10.4324/9781003489665-2

loneliness. Former Surgeon General Vivek Murthy (2025) speaks to these concerns:

> This is a moment where we find ourselves at a time of great pain, and division and polarization where every institution needs to be asking itself, 'What can I do to rebuild the fundamentals of community because that is the foundation of health, happiness and fulfillment.' (0:59–1:16)

Many factors related to our planetary crisis contribute to this isolation and loneliness, and I will explore each in this book. Here, I highlight the isolating effects that come from clashing beliefs, as well as the potential for synergy that can come from approaching differences with openness and curiosity. This in no way implies that we should compromise on our beliefs. However, the first step is to understand them, and that is the focus of this chapter.

UNDERSTANDING THE SPECTRUM OF CLIMATE CHANGE RESPONSES IS ESSENTIAL TO FIGHTING ISOLATING POLARIZATION

Findings cited in *Climate Change in the American Mind: Beliefs and Attitudes* (Leiserowitz et al., 2024), a joint report by the Yale Program on Climate Change Communication and George Mason University's Center for Climate Change Communication, reveal that the way Americans think about climate change is more fluid and multifaceted than the denier versus activist binary suggests. For example, findings from the national survey show that 62% of Americans say they "never" or "rarely" discuss concerns about global warming with friends and family, even though 64% of Americans report feeling at least "somewhat worried" about it and 28% report feeling "very worried."

Some Americans who are concerned about the problem are ambivalent about solutions. For example, 65% say they believe that climate change affects them personally, but only 31% say they have rewarded companies that take action to address global warming. Feeling brave enough to have casual conversations about such an abstract, vast, and depressing topic is tough.

As I have learned more about how psychology and environmental advocacy intersect, I have seen how emotions surrounding our global environmental crisis present themselves in psychotherapy sessions in subtle and yet profound ways. Relationships become strained, for example, when one partner feels frustrated over the other's insistence on sustainable travel options. In my own practice, some clients tell me that they want help with their anxiety, depression, and attention-deficit/hyperactivity disorder (ADHD) but the chaos of existential crises underpins much of the "noise" that rattles them in their daily lives. I have seen this pattern unfold in my personal life as well, and I will discuss these experiences in this book along with examples from my clinical work.

I hope this book will show that you do not have to subscribe to labels like "activist" or "climate scientist," to be engaged in fighting for an environmentally clean future, and you don't need to be overwhelmed by eco-distress to benefit from managing the difficult emotions environmental crises bring up for you. In psychology, labels help us understand people, but their downside is that they can encourage polarization.

Client–therapist interactions are complex when it comes to this issue. In a recent survey, one team of researchers found that nearly half of mental health providers don't feel prepared to address climate change concerns with their clients (Reese et al., 2023). Perhaps sensing a disconnect between therapy and the politics of climate change, many clients are also hesitant to express environmental distress. In the age of needing quick solutions to every conundrum, the possibility of concrete answers to existential ones seems forlorn.

This brings me to an important misconception about cognitive behavioral therapy (CBT) that I want to debunk immediately, and one I return to throughout the book. CBT for eco-distress doesn't assume distress about climate change is irrational. The practice of not assuaging existential fears isn't new in CBT. For example, when I was in graduate school, a supervisor emphasized the importance of being honest with clients who feared the afterlife. The appropriate response is to admit that we don't know for sure what it is like to be dead.

Regardless of therapeutic orientation, a therapist with strong clinical acumen aims to validate their client's concerns, however abstract and existential they may be. They help the client cultivate acceptance and build agency in addressing problems to the best of their ability. This includes teaching the client how to manage emotions that interfere with goal-directed behavior, such as climate activism. It also includes strengthening interpersonal skills so that they are empowered to build their own support network independent of the therapist, as well as see themselves as contributing to their community. That work is not going to get done if we spend the hour telling the client that everything is going to be fine or that they shouldn't worry.

CBT for eco-distress is not about telling the client they are being irrational, and yet many, including other clinicians trained in different therapeutic orientations, automatically assume this to be the case. Mental health professionals generally agree that doing something to feel like you are part of the solution, like composting, is therapeutic and that the psychological benefits of taking action are stronger when people join with environmental groups (Schwartz et al., 2022). However, CBT can take this one step further by helping the client acquire skills to sustain their efforts, including interpersonal skills to deal with things like setting boundaries with other activists and emotion regulation skills such as knowing when to take a break.

It is imperative for mental health professionals to be wiser role models, as we know all too well that problems, including mental health ones, are not only about the individuals alone, but also about how they function and interact with their social network and broader community. Before we attempt to mend dysfunctional relationships within that system, though, we need to highlight the emotional hurt being caused by widespread disunity in general.

UNDERSTANDING DISUNITY AND ITS EFFECTS ON MENTAL HEALTH

The results of the 2025 World Happiness Report (WHR) (Helliwell et al., 2025) suggest that polarization is making us miserable. The WHR details annual findings from an international research project based on people's ratings of their subjective well-being. Both the 2024 and 2025 reports made the headlines of *The New York Times* (Deb, 2024, March 20; Deb, 2025, March 20), because in 2024, the United States failed to rank among the top 20 happiest countries for the first time since the report started in 2012. The 2025 report showed even more dour results, with the United States ranking 24. Further analysis of both reports revealed that younger people (under 30-years-old) were bringing down the ranking.

Why are young adults so unhappy? The 2025 WHR (Helliwell et al., 2025) explains that income is a critical factor, but it is clearly not the only one. Loneliness is another sticking point. Although loneliness is not caused entirely by global warming, the emotional effects of the climate crisis stand to exacerbate it, as well as hold people back from doing more for the planet. For example, the political dissension surrounding the climate crisis (i.e. Kennedy & Tyson, 2024) stands to worsen our loneliness and isolation. Social media can silo people into camps of dissenting beliefs. According to the 2026 WHR (Helliwell et al., 2026), social media diminishes interpersonal and institutional trust, which in turn hurts well-being.

Job prospects can also influence one's trust in social institutions, and research shows that loneliness is both driven by and contributes to job insecurity and financial uncertainty. People whose skills are not in demand are more likely to migrate across state lines, where they might lack familial support and established friendships (Macaluso, 2025). In turn, loneliness could encourage isolation, which decreases opportunities to strengthen one's position in the job market (Bryan et al., 2024). The depressed mood that accompanies loneliness might also affect productivity, including time and energy to devote to pro-environmental action. Therefore, like a virus that spreads throughout its host, loneliness leads to ignoring pressing problems like climate change, and these problems in turn worsen loneliness.

Jade Song is a young novelist (27-years-old at the time) cited in the 2024 *New York Times* piece about the WHR (para. 12–13) who comments that young adulthood comes with existential concerns, specifically in regard to the lack of control one has over global predicaments, such as climate change, and the foreshortened sense of future prospects, like having children. She states, "'Even if you're going to protests or paying your rent and bills all the time, it's so difficult, especially now, to break free from how you're living your life when you realize how little impact your actions actually have on a broader level'" (Deb, 2024, para. 12–13).

Despite research showing that *collective action* (Hamann et al., 2024), such as involvement in advocacy groups, is the antidote to existential angst, surpassing benefit from individual actions like taking shorter showers, Song's comment begs the question: Is collective action even enough? Research suggests that many Americans aren't so sure.

For example, one Pew Research Study (Tyson & Kennedy, 2023) found that most Americans don't believe that climate activism, such as protests, rallies, and volunteering, is particularly effective at pushing elected officials to act on their concerns. While this finding is skewed by political party affiliation – for example democratic liberals appear to be relatively more optimistic with 49% saying these efforts are somewhat helpful and 16% saying they are extremely or very active, overall, these responses fail to convey resounding confidence in the power of activism despite history showing that it can be an effective vehicle for change.

The results of a different Pew Research Study (Kennedy & Tyson, 2024) revealed that Americans are split on their opinions about whether climate policies are affecting the US economy, with those who identify as Democrats leaning more toward believing such policies are helping and those who identify as Republican leaning more toward believing they are hurting the economy. The study also found that eight out of ten Americans feel frustrated about the country's political dissension in climate change. Other emotions Americans report experiencing in response to climate change and policies surrounding it include sadness about what is happening to the Earth, anxiety about the future, motivation to do more to address climate change, suspicion of the groups and people pushing for action, confusion about information presented to the public, optimism that we can address climate change, and annoyance that there is so much attention on the issue.

The gamut of emotions is as wide as the potential beliefs and attitudes with which they might be associated. While endorsements such as "Motivated to do more to address climate change" (Kennedy & Tyson, 2024) are clearly pro-environmental, others, such as "Annoyed there is so much attention on this issue," are clearly not. Yet other responses give us little information about the behavioral response with which they are associated. For example, optimism, while not the same as hope, can fuel perseverance of activism in the face of adversity, or it can enable passivity with the assumption that others will fix the problem.

Understanding the complexities of responses to the climate crisis requires going beyond binaries such as liberals vs. conservatives or Democrats versus Republicans to consider the various behavioral responses that the idea of climate change might instigate. Below is my attempt to organize these responses based on what I've learned from research studies, scholars, clinical and personal experiences, into the following response types: **deniers, burrowers, preppers, victims, activists, academics, and strivers**.

I caution that these response types are not meant to represent personality types. They are also not meant to be generalizations or stereotypes, which would have the opposite of the intended effect, serving to disunify rather than strengthen collective action.

Additionally, these response types are fluid and dynamic. For instance, when I research the risk of flooding in my neighborhood, I am engaging in prepper-type behavior for averting the consequences of disaster. During any given period of time, I might also be planning a lecture about the mental health effects of climate change (academic response) or attending a protest (activist response). Personality, situational context, personal values, culture and various other demographic variables shape our behavioral tendencies, and the confluence of these factors leads us to shift among the various roles described here. Still, understanding these general responses will help us understand how different responses can lead to clashes, but, if met with openness and curiosity, they can also lead to richer collaborations.

Climate Change Deniers

Data from the national Yale Climate Survey suggests that climate deniers are in the minority, as 72% of Americans believe that climate change is happening (Leiserowitz et al., 2024). Yet tolerating climate denialists is no easy task considering that they may attack people who are not on the same side. The intensity of these reactions can be jolting. I know because I was on the receiving end of it not too long ago, when I wrote a blog post about the psychology of climate change (Keluskar, 2023). In the post, I discuss how self-consciousness about environmental behavior, such as feeling weird for bringing non-disposable Tupperware to a restaurant and piling leftovers into it, is hurting the climate cause, and that being more confident about engaging in advocacy can help assuage eco-distress.

Soon after I published this piece, I stumbled upon a vilifying response post from a climate denier's website, wattsupwiththat.com, entitled: "Psychology

Today: Climate Activism Will Set You Free from Anxiety" (Worrall, 2023). Here are some notable quotes from the criticism that follows a copy and paste of my article and name. Much of the criticism consisted of biblical references I frankly couldn't comprehend, and so, while I cite only an excerpt here, I provide the link to the entire piece as well:

> There is something disturbingly medieval about Jennifer's prescription for climate change. The Children's Crusade, a 13th century attempt to convert the heathen, after Saldin turned the tide against the Crusader states in the 12th century, was composed of anxious children who felt their parents had failed God.
>
> (https://wattsupwiththat.com/2023/07/07/psychology-today-climate-activism-will-set-you-free-from-anxiety/)

The author then updated the post to reflect possible inaccuracies in the biblical story due to "historical controversy."

Not all climate deniers have the same disdain for me as Worrall and his followers on wattsupwiththat.com, one of whom said that people shouldn't let their children near me. In fact, several of my psychotherapy clients have also been deniers. They were lovely people who developed a warm and trusting relationship with me, even after they realized that I am an environmental advocate. Rather, they reacted to this realization with mild trepidation, fearing that I would judge them and treat them differently, which of course I didn't.

They were also lonely, and anxious, presenting to sessions with concerns that they would be attacked for their views and that climate change believers would strip them of the means to earn a living and feed their families. Even though our views drastically differed, we both feared the breakdown of the natural order.

Climate deniers, including the ones on wattsupwiththat.com who had no qualms about airing their vitriol toward me, can be scary in fight mode, but in everyday life, they might be your accountants, lawyers, teachers, or the patient school bus driver who waits for your kid when they are running a minute late.

What we need to understand, and have compassion for, is that climate denialism can be entrenched in fear and isolation. By showing curiosity and creativity, we have the power to find some sort of common ground. This common ground might mean finding ways to connect outside of environmentalism that simultaneously achieve environmental goals. Recently, I praised electric cars to a conservative acquaintance and emphasized how joyful it is to never have to pump gas and to be able to conveniently fuel up my vehicle from the charger we installed in our garage.

In other cases, you can find common concerns that are environmentally rooted, although not specific to climate change. Later in this chapter,

I highlight a news story about the political efforts of both democrats and republicans who have been affected by forever chemicals.

We also need to understand some of climate deniers' concern about transitioning from fossil fuel industries to cleaner ones. The 2024 World Energy Employment Report (International Energy Agency, 2024) showed increased green energy employment globally, as well as growth in green jobs outpacing fossil fuel energy sector positions. However, certifications relevant for energy sector positions have not paralleled the growth of green job availability. In other words, green jobs are more plentiful, but the availability of qualified workers is lagging. According to the 2025 World Energy Employment Report (International Energy Agency, 2025), employment in the electricity sector, such as solar and nuclear power, has risen by 3.9 million and now represents nearly three-quarters of all energy job additions. However, a shortage of skilled workers has persisted.

Additionally, automation has made it possible for the United States to remain the second most productive country without requiring us to work harder, and green energy requires fewer workers than the fossil fuel industry (Communication with D. Wrenn, March 25, 2025). While opportunities might exist for coal miners to pursue higher education, shifts in the market have favored health care. In a similar vein, economic and social theorist Aaron Benanev notes in an editorial he wrote for *The New York Times*: "The move from manufacturing to services is happening across the G20 [a forum of the largest economies in the world, World Economic Forum, 2025], dragging down growth rates nearly everywhere" (Benanev, 2025, para. 8).

All of this suggests two things: One, that to encourage collective action, we also need to address poverty, which is preventing people from being able to engage in sustainable behavior. We also need to think outside of the box to understand how health care and other human services, which are driving the economy, can help people create lifestyles that don't rely exclusively on technology to save us from environmental disasters.

As mental health professionals, we can also help people change their relationship with the resources they use, including energy consumption. In his talk *Power Shift: Shattering Illusions about the Energy Transition and Our Future,* Nate Hagens talks about the "Great Simplification" as a cultural revolution that will be the greatest event in human history (Hagens, 2025). It is the idea that economic growth will no longer be an option at some point, and we will need to slow down in order to sustain life on this planet. This level of change might sound scary, but it is necessary and can be joyful if we transition to lifestyle changes in a timely manner. For example, we can focus on time in nature over fancy vacations.

First, though, we need to agree on our goals, and polarization is making this challenging.

A 2024 study of posts on the social media site X (formerly known as Twitter) (Gounaridis & Newell, 2024) found that two communities, climate deniers and believers, existed in vacuums, mostly sharing information within their own community. This resembles what Monica Guzmán (2022) refers to as siloing (pp. 31–50) or the act of clustering around those whom one sees as being like them. In her book *I Never Thought About It That Way*, Guzmán emphasizes that when there is a lack of diversity in viewpoints, the threat of the other becomes more ominous. She states, "When the people I talk to get nervous about crossing the political divide, it's usually because they fear or loathe what they expect to find there" (p. 43). These assumptions dehumanize the other side and perceive it as a conglomerate of depravity rather than difference.

The politicization of the climate crisis has allowed the festering of cultish behavior, not the least of which concerns climate change denial propaganda. The 2014 documentary *Merchants of Doubt* (Kenner, 2014) portrays how disinformation was spewed by government-backed science representatives with the intention of sowing doubt about the anthropogenic roots of the climate crisis. Disturbing parallels were drawn to past disinformation campaigns fueled by the tobacco industry, which stoked public doubt about the health risks of smoking.

Personality also plays a role in climate denialism. Specifically, climate denial is associated with rigid thinking (e.g. Scheffer et al., 2022) leading people to be ill-equipped for unifying with those seen as deviating from their ideology. They are also more likely to be closed off from new learning that could lead to improved health and well-being.

Cognitive behavioral tools aimed at increasing flexibility can help climate deniers tolerate those who differ from them and attempt to find common ground with them. While deniers are unlikely to want help with their denialism, they have shared with me their distrust and discomfort with being in the mere presence of environmental advocates. This is where a therapist could work with them on improving psychological flexibility, which could help them feel more comfortable and possibly set the stage for openness to others' viewpoints.

There are ways of subtly shifting our culture's relationship with resources without imposing our beliefs on others. The late, legendary conservationist Jane Goodall spoke of telling stories as a more effective way of shifting denialism than spewing scientific findings (Mendoza & Knudson, 2025). Mor Keshet MPS LCAT, an integrative eco-art therapist based in Huntington, New York, shared that she engages in "gentle but very clear modeling" of sustainable behavior. For example, when people compliment the décor of her office, she shares that she purchased used items from Facebook Marketplace (personal communication, July 28, 2025).

Climate deniers are not the only ones who might feel that they exist on an isolated island. Research (Sparkman et al., 2022) highlights that Americans underestimate how many people are concerned about climate change. People's predictions of climate concern ranged from 37 to 43% on average, when data reveals that in reality, 66–80% of Americans support policies aimed at lessening the impact of, or mitigating, climate change. This means that those who are worried about climate change might believe that they are alone in their concern.

This finding is refreshing in the context of a crisis that can feel incredibly invalidating. Understandably, the possibility of our extinction isn't exactly a winning dinner party conversation topic. Even when I've tried to chat about it nonchalantly, such as saying, "Geez with climate change, we are going to get crazier weather," on an 80-degree day in April, it has inevitably been met with an anxious grimace from the other person, followed by a prompt change of subject. Then again, on the occasions when people in the community bring it up to me in casual conversation, I'm so shocked that someone else brought it up I freeze for a moment before responding, likely unintentionally conveying discomfort myself.

When we lack safe spaces for climate change discussion, it can be easy to believe that nobody else thinks it's important or concerning. Isolation, hopelessness, and angst ensue. In some cases, it is easier to distract oneself from all of this and act like it isn't happening. Enter, the burrowers.

BURROWERS

Burrowing is how I describe sticking one's head in the sand to avoid the overwhelming implications of human-caused climate change. Some sources have referred to burrowing as another form of denialism (Haltinner & Sarathchandra, 2018). Burrowers attribute climate skepticism and denialism to actively seeking a safer narrative in order to avoid unsettling truths.

According to a *New York Times* piece about why people fail to notice horrors around them (Sharot & Sunstein, 2024), "When threats start small and increase gradually, they end up eliciting a weaker emotional reaction, less resistance and more acceptance than they would otherwise. Larger and larger horrors play out in broad daylight – taken for granted, seen as ordinary" (para. 3). While the article, written by professors of neuroscience and law, wasn't specifically about climate change, the topic seems apt. It reminds me of when my then six-year-old son asked, "Mommy what happens to our trash after the garbage men take it away?" Most adults, even though they know the answer, don't care to think about landfills and incinerators releasing toxic fumes into the air every time they take out the trash.

In this sense, burrowing is ingrained into the fabric of our society, and we are both victims and perpetrators of it. As with other forms of victimhood, this doesn't doom us to victim mentality, nor does it relieve us of our obligations to care for the earth our descendants will inherit.

However, categorizing burrowing as pure denialism doesn't seem to fit either. As the Yale Climate Survey showed, most Americans are concerned about climate change and yet go about their daily business as if it's not actually happening. This response comes from assumptions, such as: "There is nothing I can do to stop climate change," "The powers that be will take care of it," or "Climate change is a far-off rather than a present-day problem, so advances alone (without necessary changes in behavior) will resolve the crisis."

It might feel surprising that the public is not raging or panicking at the gates of the White House as they surely would be if they believed an extinction level event were at their doorstep. Like the climate deniers, burrowers fail to prioritize environmental policy, but they won't necessarily deny that climate change is real or even driven by human factors. They might even sigh and shake their heads if you bring up the subject, but then quickly change it to how bad traffic has been lately.

Another distinction between burrowing and outright climate change denial is that burrowing is not always unhealthy. It is something we all engage in, and I would argue, must engage in to an extent in order to preserve sanity.

Refusing to occasionally burrow on a small scale can fuel burnout. For example, some of my clients have felt guilty about not spending enough time watching footage of horrific world events despite feeling exhausted and sleep deprived.

While my head isn't buried in the sand, I can't imagine I would be able to function well if I didn't at least sometimes put climate change on the mental back burner. This is in line with the media diet that mental health professionals recommend, which entails both putting limits on doomscrolling scary news about climate change and mixing up the hard truth with more uplifting truths about work that is being done to mitigate the effects of climate change (e.g. Crooks, 2024).

But I also can't live in the sand 24-7. Climate change has left an indelible print in the backdrop of my mind. It doesn't exactly take over when I am engrossed in work, housework, and childcare, but, given I'm reminded of it whenever a faucet or light is left on, or when I open my news app to learn of yet another incident of catastrophic flooding that destroyed people's lives, it never quite leaves the room either.

CREATIVITY CAN HELP US BREAK OUT OF BURROWING

The moments we allow ourselves to sit with unsettling reality leave room for inspirational waves of creativity. Take, for instance, the time when I had saved candy wrappers and my then nine-year-old daughter wanted to use them to make a "candy mom" (see Figure 1.1), in which she glued the wrappers to create the shape of a person.

Mor Keshet, MPS, LCAT, an integrative eco-art therapist and author of the paper, *Integrative Eco Art Therapy: Addressing Humanity's Fractured Attachment to Nature* (2025), describes Integrative Eco Art Therapy (IEAT) as an approach that encourages psychological healing through finding meaning and purpose. She also describes it as self-actualizing, or fulfilling one's potential, as well as eco-centric, or rooted in ecological themes. Mor helps her clients mend unhealthy attachment and cope with past trauma by guiding them in creating art with nature-based themes or using materials from nature like twigs and leaves in their creations.

Figure 1.1 *"Candy Mom" by Skylar Keluskar.*

Photograph by the author.

When I attended a workshop led by Mor in the fall of 2023, she provided me with earth clay which is sustainably sourced art material. She gave attendees free range over creating whatever we were inspired to make. I molded a bird's nest with eggs. Mor guided me in interpreting my creation as representing a maternal role. What also piqued my interest was a conversation about my bird's nest nearly two years later. At the time, I was overwhelmed by the tasks and projects I was juggling and mentioned to Mor that the eggs represented my ongoing struggle to keep my projects from failing. Mor commented that sometimes we need to accept that some eggs will break. The conversations that emanated from my simple creation show how art can be another medium for confronting struggles, such as perfectionism. Spending time in nature teaches us that nature's beauty and wisdom shines through not only in spite of but because of imperfection.

Creative approaches like those Mor uses have potential to awaken burrowers to move toward hope rather than futility. Rather than condemn burrowers, we need to gently show them the potential for empowerment that can come from less than perfect acts of sustainability.

Climate change is not your fault. But it is your responsibility to address it. At the same time, you can't do it alone.

Skills for increasing psychological flexibility can be applied toward making this kind of collective action happen.

PREPPERS

Prepping is the practice of "anticipating and adapting to impending conditions of calamity, ranging from crises to extinction-level events" (Garrett, 2021, p. 401).

As described in a July 2023 *New York Times* article entitled "As Climate Shocks Multiply, Designers Seek Holy Grail: Disaster-Proof Homes" by Christopher Flavelle (2023), there is a trending interest in custom-built geodesic dome homes built to withstand hurricane-force winds. Other weather-resistant upgrades include steel and concrete homes. These additions cost about 10% more than conventional construction.

What about those who can't afford basic housing and electricity, let alone central air conditioning, and even more unrealistically, weather-resistant custom-made homes? I think about a heartbreaking story about deadly floods from a monsoon in Pakistan that swept away entire villages (Ur-Rehman & Peltier, 2025). A village was unable to receive weather alerts warning of severe weather due to poor internet access. Like many other aspects of the climate crisis, prepping brings to light disparities that natural disasters and other effects of climate change will only worsen.

For those who have access to streaming services, the word "prepper" might bring to mind the reality television series *Doomsday Preppers* (Madison et al.), which aired on *National Geographic* from 2012 to 2014. The show featured survivalists who took extreme measures to prepare for hypothetical apocalyptic events, whether they be natural disasters or financial collapse. Think bomb shelters lined with years' worth of preserved foods. The participants were also subject to mockery. The show made it a point to emphasize the unlikelihood of the anticipated events for which these individuals make tremendous efforts and sacrifices.

The participants portrayed the extremes of prepping. But most of us prep in smaller ways than would win the appeal of the big screen. Prepping, to some extent, is part of human nature, particularly in response to or in anticipation of scarcity (Smith & Thomas, 2021). This trend was commonplace, for instance, in the face of shortages that transpired during the Covid-19 pandemic when people hoarded toilet paper.

Young adults I see in my clinical practice prep in anticipation of a withering social safety net, compounded by perpetual student loan debt. In the face of climate change, they are uncertain about whether resources will be available in the future, and they feel pressured to create a nest egg more quickly than is realistically possible for most people who are just starting their careers. Relatedly, they believe they must acquire wealth first before they can focus on philanthropic efforts aimed at sustainability.

On the one hand, we can see prepping as a normal response to the prospect of ecosystems and civilization collapsing, which scientists predict in the event of unmitigated global warming (Domonoske, 2018). On the other hand, extreme prepping captures a mental state steeped in doom and dread. As opposed to denialists, preppers respond to these emotions by acting as if the worst-case scenario will play out (Garrett, 2021). The problem with this response is that an over-reliance on prepping might lead to an underemphasis on energy and resources that can be spent on taking action to prevent the worst outcomes that could come from climate change.

Preppers exclusively depend on *adaptation* or altering their living situation to enhance chances of survival in the face of environmental threats, a response that is insufficient according to the Fifth National Climate Assessment (Chapters 31 and 32). More effort must also be devoted to *mitigation* or combatting the main player in the human-caused climate crisis: The fossil fuel industry.

In 2023, acclaimed climate scientist and activist Peter Kalmus visited the Possibility Alliance, a fossil-fuel free homestead in Maine (Bush, 2023). He commented that he enjoyed the low-energy lifestyle that is missing from modern life. At the same time, Kalmus argues that no place is safe from climate change. He had moved to the Northwest thinking of this region as a climate haven, only to experience a three-day heat wave.

In the face of Elon Musk's goal of colonizing Mars, Kalmus has spread the word that "there are no other planets we can live on or can currently get to. And Mars totally sucks. It's cold. It's a desert. There's no water. You're bombarded by radiation that would give you cancer …." (WIRED, 2022, Chapter 2, 0:22).

The elite prepping phenomenon can also take the form of secure, stocked basement abodes (Garrett, 2021). For some, a prepping plan represents a solution to catastrophe and simultaneously a respite from the pressures of the world as well as an opportunity to gain a renewed sense of purpose.

During the height of the Covid-19 pandemic and mandated quarantine, the tragedy of lives lost and disruption stood side by side with fresh ideas for creating resources and facilitating remote work. For example, by working from home, I was able to spend more time with my family. I felt inspired to write and create projects. Of course, the allure of a reset on life is a delusion in the face of an apocalyptic situation, but the allure of a life away from the pressures of living within modern society is well noted.

As mentioned above, prepping conceptually parallels the notion of adaptation to the effects of global warming. As sea levels rise, we will need to continue to build homes that are elevated to prevent destruction from flooding. Inevitably, though, an extreme prepping mentality, without sustainable action, is selfish and overly optimistic. Humans are not built to survive alone, and in modern civilization we have forgotten how hard it is to grow a strawberry larger than my child's thumb nail, or to treat water so that it is not contaminated with harmful pathogens.

From the perspective of Radically Open Dialectical Behavior Therapy (RO DBT) practitioners, the cognitive process behind extreme prepping might be categorized as fatalistic thinking (Lynch, 2018a), or the "escape artist" of the mind that throws in the towel, often selfishly. Fatalistic thinking also exacerbates isolation, loneliness, and hopelessness.

Climate Tech Innovation (Seibert, 2025), such as technology for capturing carbon dioxide from the atmosphere, represents part of a proactive solution to the climate crisis, but even this takes on a prepper mentality when it is not paired with a shift in societal values away from materialism and toward sustainability. In a 2019 *New York Times* Piece, "How Big Business is Hedging Against the Apocalypse," investors are seeking alternative, less expensive power sources, such as wind farming. It is another reminder that many proposed solutions to the climate crisis entail embracing the same spirit of capitalism that is at its root cause. Any fix that bloats the pockets of billionaires understandably raises an eyebrow.

Climate Tech Innovation

In *The 100% Solution: A Plan for Solving Climate Change* (2020), Solomon Goldstein-Rose, a former Massachusetts representative with an educational background in engineering and public policy, joins the consensus of scientists confirming that cutting emissions is a necessary, but not sufficient, solution to the climate crisis. We will also need innovation to prevent further damage, as well as to help the next few generations cope with effects that are now inevitable. Goldstein-Rose is concerned that political rhetoric focuses on one solution, such as 100% renewable electricity as opposed to more comprehensive plans.

Technology for addressing climate change has already been employed, as evidenced by wind farms, solar paneled homes, and even the beginnings of more ambitious efforts such as Orca, the first large-scale carbon-dioxide removal plant in Iceland (https://climeworks.com/plant-orca). Other proposed technology for "curing climate change" is far from ready for deployment, such as building giant sun shields in space (Buckley, 2024).

Professor David Keith, a physicist at the Harvard School of Engineering, has spoken to the public through videos and podcast interviews, about the potential of a controversial technology called solar geoengineering (Haskett, 2023), which encompasses techniques for making earth more reflective, such as making the ground whiter or putting tiny particles into the air. Critics of these approaches argue that such interventions might present yet unknown threats to humanity. Furthermore, it is feared that they will distract the public away from the importance of ending the fossil fuel industry (e.g. behavior change).

In reality, both approaches are needed, and we can't escape the formidable truth: The need to re-invent social norms. According to Aaron Benanev, an economic historian and author, strengthening the United States and global economy can be achieved by levying higher taxes on the rich and redistributing income to less wealthy individuals, but goals would need to shift away from becoming richer to stiving for stability (2025, paras 17, 18).

All of this is to say that it is not technological advancement alone that is unrealistic, but rather the challenge lies in aligning impressive scientific feats with a stance of open-mindedness and cooperation over materialism. This brings us to the next response type in the realm of academia.

ACADEMICS

A 2018 qualitative analysis (Hoggett and Randall) showed that climate scientists tend to communicate their research in logical, unemotional language. The authors of the study argue that this has been detrimental to convincing the public of the urgency of the climate crisis. The authors also believe that competitiveness, or the pressure to earn funding and recognition through grants and publications, pulls scientists away from fighting for change.

According to a 2024 Pew Research Study (Kennedy & Tyson, 2024), 89% of Americans saw research scientists as intelligent while only 45% endorsed seeing them as good communicators.

While it is important for data to be translated into terms that are relatable to the public, a translation is only effective if there is a willing learner, on the other end, and unfortunately this is all too often not the case when it comes to climate science. Adam McKay's (2021) film *Don't Look Up* satirically depicts the disconnect between scientific logic and how people who don't want to believe it might respond. An astronomer and his student (played by Leonardo DiCaprio and Jennifer Lawrence, respectively) attempt to convince the US President (Meryl Streep) that they have predicted an asteroid will collide into Earth, with absolute certainty. The president's son and second-in-command (played by Jonah Hill) questions this claim given that the predictions were "99.87%" to be exact. "Oh great, he says, well it's not going to be 100%!" (19:55). When the scientists confirm that the asteroid hitting the earth is a certainty, the President emphasizes that we can't tell people they are going to die. The point here is that academics can tell truths that do not resonate with the public either because their writing isn't sexy enough or because it is seen as Cassandra-like.

Lacking the art of persuasion is not academics' only problem. There's also good reason why people aren't flocking to scientific journal articles despite these readings being relatively more accessible than in the past. In an episode (June 13, 2021) of the podcast *No Stupid Questions* with Angela Duckworth and Stephen Dubner, Duckworth and Dubner answer a listener's question, "Why is academic writing so bad?" The hosts present both sides of the argument for erudite journal articles. On the one hand, scientists need to be able to communicate in a common language, and therefore the use of jargon, as with any other profession, becomes relevant, if not essential. On the other hand, not every scientist deems it worth their time to develop writing as a skill, let alone become an influencer. A promising solution is the idea of journalists being trained to write in a way that catches the public interest, teaming up with researchers. Britt Wray, journalist and author of *Generation Dread: Finding Purpose in an Age of Climate Crisis* (2022), has co-written research papers on emotional responses to climate change (e.g. Lewandowski et al., 2024) and also has a Substack where she shares research-informed posts about the effect of climate change on mental health in language that is more digestible to laypeople.

Some climate scientists such as Peter Kalmus have stepped up to the challenge of advocating for climate scientists to get off the sidelines. Kalmus has been arrested twice for protesting, including one occasion when he shackled himself to a JP Morgan Chase building in Los Angeles (Bush, 2023).

It is understandable, though, that academics hesitate to get involved in activism given the risks to their careers. For example, Gianluca Grimaldi lost his position at the Kiel Institute in Germany in October 2023 for refusing to fly home from a fieldwork project (Buckley, 2023). Another environmental scientist, Michael Mann who at the time was at Pennsylvania State University (he is now on faculty at University of Pennsylvania) and recipient of the Climate Champion Award, was falsely accused by a blogger of manipulating data (Simon, 2024). He subsequently won a defamation lawsuit.

Addressing climate change through one's career also makes sense given that for many, this is the arena of one's belief in their competence. However, getting around the wall of academia can feel daunting. When I joined a special interest group for CBT and climate change, the conversation seemed to focus more on publications than practical discussions for how clinical psychology as a discipline will confront environmental devastation.

We will lay this section to rest by emphasizing the incredible value of labor that scientists have put into shedding light on the magnitude of the climate crisis as well as potential solutions. Their importance can't be understated. However, just as within the other climate response types discussed, there are differences in opinion among academics that will need to be addressed.

Take, for instance, a *New York Times* piece (Tabuchi, 2024) about two climate researchers who are friends, Rebecca Grekin and Yannai Kashtan, and yet hold fundamental differences in how they go about their research. One such difference is the question of whether they should allow the fossil fuel industry to fund their work. For example, when Grekin wanted to research more efficient air conditioning and heating systems, Exxon provided her with an entire building to use. While Grekin is able to complete her research because of this funding, Exxon uses the relationship as a way of boasting their support of young climate researchers, despite clearly also profiting from polluting the environment.

Grekin had to deal with criticism that the research was biased due to the source of funding, which researchers disclose when publishing their data (Tabuchi, 2024). Nevertheless, her research has led to limiting the climate impact of 40 universities' food services. Kashtan is wary of research about far-reaching technological solutions that might distract from the importance of using fewer fossil fuels.

Despite their differences, they are also friends, stepping in to teach each other's classes (Tabuchi, 2024). More importantly, they agreed to guidelines that require researchers to allow funding from the fossil fuel industry only if the company has a plan for transitioning away from fossil fuel use to renewable power. While Kashtan doubts that this measure will be effective, their climate work emphasizes collaboration despite lacking unilateral agreement for how to solve the climate crisis. In a way, their respect for their differences leads to a diversity of responses which reflect what we ultimately will likely need: a multifaceted solution, not just one.

Whether it is through writing or teaching, academics can disseminate knowledge informing activism. The next section focuses on activism as a response to climate change.

ACTIVISTS

The *Britannica Dictionary* defines an activist as a person who uses or supports strong actions in order to bring about political or social change (2025).

Which actions constitute activism can be subjective and, at times, a point of contention among environmental advocates. Going off the above definition, am I an activist for following a boilerplate script in an action call to New York Governor Kathy Hochel's office? Some might consider this *slacktivism* (Merriam-Webster Online, 2025), which requires little commitment or effort, presumably having less influence on positive change, although some experts argue that it can have subtle as well as accumulative effects on public opinion (Fisher, 2020).

Climate scientist and activist Peter Kalmus urges people to be more active in their activism. According to his post on X on July 21, 2023: "If you haven't done any sort of climate activism yet, please do (Recycling, driving an EV, getting solar panels, composting, etc. – while OK things – are not climate activism)."

I've had a complicated relationship with activism throughout my adult years. When I was in college in the early 2000s, a friend convinced me to join the Student Global AIDS Campaign. I enjoyed time under her wing and had adventures like a road trip to Washington DC for a student rally. We protested against big pharma blocking lower income countries from producing cheaper, generic versions of life-saving Antiretroviral treatments.

However, as time passed, I grew more aware of my inadequacy at activism. Student activists were on track for careers in law, political science, and medicine. They were quick to sniff out my lack of confidence and knowledge, and to send me to the back room to stuff envelopes rather than do anything that felt meaningful. On top of that, as schoolwork piled up, my focus shifted to graduate school applications, and at the time, activism wasn't an attention-grabber for clinical psychology doctoral programs. I didn't have time for extracurriculars that wouldn't help me further my career, let alone for working with activists who saw me as useless.

I took a break from activism after college, but I felt compelled to try again after I had children and my concerns about the climate crisis escalated. I started by reaching out to local agencies. They didn't respond.

Finally, I found one and learned that the team leader for my region, whom I will call Sally, was the sole member. That didn't matter to me. In fact, the idea of being at the forefront of growing a team excited me. This time, I brought a different attitude to it than my less secure college self almost 20 years prior. I didn't need the tasks assigned to me to be important because I had the perspective to realize that even small actions like handing out flyers could contribute to a larger cause.

At first, that was good enough. I helped out with tabling at local fairs, carrying stacks of sign-up sheets and handing out postcards about a bill that would mandate new construction to be powered by clean energy. I unsuccessfully tried to recruit more members, but still, I pressed on.

Sally and I would have long chats about how energy was managed in our area. When I summarized what she said to make sure I understood correctly, she told me I wasn't understanding the concepts. She started making comments like, "I wish we still had [name of someone who left the team] here. That one knew her stuff!" Initially, I tolerated these jabs because I accepted that I could only do the best I could, and showing up was what mattered. Ultimately, though, I left the organization, but not before regrettably snapping at Sally for bossing me around.

In hindsight, it would have been more effective to talk to Sally about how I was feeling before I lost my cool with her. After reflecting on this for months, I realized that I hadn't wanted to acknowledge feeling hurt by her petty remarks. Simply put, my concerns seemed trivial in light of the enormity of what brought us to those tables to begin with: The onerous weight of earth's break-down. Unfortunately, this self-invalidation took me farther away from taking any action in resolving my conflicts with Sally.

Here is my wish list for the leader of any organization I join in the future:

1. *Address perfectionism.* Perfectionism, associated with rigid thinking (e.g. Goulet-Pelletier et al., 2025) tends to fuel either indignation, when things must be a certain way, or resignation when there is no use in trying because a leader's expectations cannot be met (e.g. Lynch, 2018). We need to empower everyday people to advocate for pro-environmental measures.

   Perfectionism is explored further below as well as in later chapters of this book. Ironically, it often runs counter to sustainability, such as compelling me to tear off more of a paper towel roll to hastily spot clean any imperfect smear in my kitchen, or to click on a must-have Amazon purchase to fix some flaw in my appearance, character, or intelligence. Thomas Curran, author of *The Perfection Trap: Embracing the Power of Good Enough* (2023), discusses the connection between perfectionism and environmental destruction. In contrast, "Good enough" has the power to heal our psyche as well as the ecosystems we depend on for survival. I believe we have an argument here for both letting go of pursuing perfection and embracing compassion in themselves representing forms of activism!

2. *Make things easier:* Many organizations use QR codes to reduce paper. This also removes the burden of handing out flyers, thereby opening more opportunities to authentically connect with others and to engage them in conversations about the cause you are advocating.
3. *Help people feel empowered by their own projects.* This enhances personal self-efficacy, or the belief that one has the capability to succeed at a goal, as well as participative efficacy, or the belief that one has made a meaningful contribution to the collective goal of a group (Sustainability Directory, 2025).
4. *Question what you think is easy.* People think that getting others to join a cause they are passionate about will be easy, but this is a fallacy fueled by assuming that everyone thinks like you do.

My experience has shown that getting others to join your cause is one of the most challenging tasks, period. In contrast, doing the work first is easier than trying to solicit people to join a project that hasn't launched.

Perfectionism within Pro-environmental Behavior

Importantly, perfectionism can interfere with self-care and contribute to burnout. On December 28, 2022, I sat down for a Zoom call with Sami Aaron, founder and director of The Resilient Activist, a non-profit agency dedicated to meeting the emotional needs of climate activists (https://theresilientactivist.org/).

The Resilient Activist operates on the principle that we need to help activists take care of themselves first and foremost. This became Sami's mission after her son tragically committed suicide in the early 2000s and Sami learned that his psychiatrist didn't realize that he had been a climate activist, let alone that the uphill battle of climate activism contributed to his despair.

I asked Sami if we could talk about perfectionism in activists. She shared an example of how perfectionism can lead us to lose sight of the big picture, thereby holding back progress: Balking at gift-giving during the holiday season because the resources they consume don't bode well for the environment.

> "But," Sami emphasized, "for so many people, giving gifts brings them joy. Activists need to find a way to do what brings them joy during the holiday season! Giving thoughtful, Earth-friendly, or homemade gifts can bring joy back into the celebration!"

According to Leo Goldsmith, a doctoral student at the Yale School of the Environment who studies the disproportionate effects climate change has on health in individuals in the LGBTQIA+ community, and whom I interview later in this book, joy is a form of activism (personal communication, February 25, 2025). Activism and joy can seem like oxymorons in a world that has completely invalidated climate activists. But joy *is* activism because it is what sustains work that can otherwise feel defeating. Moreover, joy that is shared with others facilitates belonging and demonstrates power in numbers. It reminds us that we need to help activists make life-sustaining choices.

Hoggett and Randall (2018) describe activists' emotional trajectory, starting with an epiphany moment, followed by immersion in learning about climate change accompanied by feelings of anxiety, anger, confusion and bewilderment, or what they term the *crisis stage*. Activists have found it helpful to cope with these emotions by cultivating agency, or a sense of having control over their circumstances, as well as maintaining a work-life balance.

Cultivating community is also helpful (Hoggett and Randall, 2018) and this might necessitate addressing perfectionism, which drives people to prioritize performance and achievement over social connection (Lynch, 2018). However, we need to also acknowledge society's lack of support for activists both in terms of recognition and financial compensation. Either you can't afford to make it your life's work outside of your actual job, or you earn peanuts for being an activist for a living.

Activists might also experience other-oriented perfectionism, defined as having unrealistic expectations of others (Curran & Hill, 2019). This might be directed toward volunteers or less committed members of an agency who don't have the capacity to commit as much time to organizing. This has the potential to fuel resentment within all involved.

RO DBT (Lynch, 2018) helps us feel more connected with others, embrace imperfection in ourselves and others, and develop insight into passive aggressive behaviors.

I teach RO DBT skills to clients who feel pressured to engage in activism beyond their comfort zone as well as to those who are put off by what they perceive as others' excessive activism, such as feeling frustrated that a friend is balking on taking a flight to join them on their vacation. One set of RO DBT skills that can be helpful in such situations surrounds Over-Controlled states of mind (Lynch, 2018), which include *Fixed Mind, Fatalistic Mind* and *Flexible Mind.* People who strive excessively might respond to adversity with Fatalistic Mind, which is characterized by urges to shut down, quit, or give up. Fixed Mind comes up when one is stuck in believing their perspective is the only correct way of responding. The key is to nurture a healthy Flexible Mind. By cultivating Flexible Mind, we recognize that slowing down and being open to others' perspectives is the healthiest and in the long term more meaningful and productive option.

Emotional obstacles confronted by activists include fears that change will not happen in their lifetime as well as valid fears for their physical safety, as protesters have been teargassed, beaten, imprisoned and even killed (e.g. Global Witness 2025).

Writing this book has enhanced my appreciation for the emotional burden that encumbers climate activists because it has forced me to immerse myself in the threat to life on earth that climate change carries. I have found myself mired in guilt for occasionally taking longer showers and also distressed when I imagine a future for my children in which they might not have enough resources to survive. Even though I had knowledge of the crisis before beginning this process, there is something to be said for eating and breathing it day in and day out and the potential this has for fueling burnout.

Cognitive behavioral therapists can borrow from interventions that climate advocacy groups have created to promote self-care among activists. To learn more about the Citizens' Climate Lobby's self-help group based on Ongo (Cadden & Wiens, 2017), I joined one with no charge incurred and no questions asked about how involved I was in activism. This was a fruitful experience that I benefited from in general, as many of Ongo's principles, such as forgiveness work and identifying emotions, were in line with healthy emotional practices. At the same time, connecting with climate activists helped me feel validated in my concerns. Importantly, among these activists, I felt no judgment, only communal warmth, in a way that I haven't been able to feel when I've tried being an activist.

Aspects of personality, such as perfectionism, as well as ideological and cultural differences can interfere with one's identification with a group, and therefore these factors will be closely examined in later chapters of this book. CBT skills can help individuals effectively communicate their needs, such as for rest and respect for boundaries, and to know when to find other groups, as there are a variety of climate activism groups that people can try out.

Activism is fueled by loss, and this loss might be traumatic (occurring in the past or unfolding in present), or pre-traumatic, which is anticipatory (Van Susteren, 2021). Activism spurred by past loss can also be anticipatory. When families of murder victims establish foundations (e.g. to fight violent crime, domestic violence, etc.) they do so to prevent others from suffering similar tragedies. Next, we explore victims and their response to climate change.

VICTIMS

How does victimhood apply to our experience of climate change and other forms of environmental devastation? The answer is, in pretty much every way imaginable. From the increased toxicity of polluted air and water to the known effects of heat on mental health, living on an earth-on-dialysis is bound to bring a flood (no pun intended) of obstacles to quality of life, including health and well-being.

The mental health effects of climate change can manifest in anticipatory angst about the environmental catastrophe that is likely to ensue in the coming decades (Sturm et al., 2021). Climate change can also exacerbate pre-existing mental health difficulties, such as by interfering with sleep, which in turn disturbs mood and behavior (communication with James McKowen, August 8, 2025). Psychiatric medications can also increase the body's sensitivity to heat, thus throwing the body's thermoregulation further off balance during a heat wave (CDC, 2024).

Victims of natural disasters are at increased risk for being the targets of crime and violence, particularly marginalized individuals such as people of color, sexual and gender minorities, and women and children (Shreve & Fordham, 2018). Victims of natural disasters stand to have damages inflicted on their sense of purpose, through loss of employment, as well as on their sense of security and identity, through loss of their home and other objects of value (Heanoy & Brown, 2024). The disruption of normalcy creates distress, particularly for children, when they are unable to go to school or see friends.

Mental health symptoms that emerge after a disaster can persist for years after the event. Linda Goldman, author of *Climate Change and Youth: Turning Grief and Anxiety into Activism* (2022), has worked directly with children affected by natural disasters. She comments on how events that adults might not prioritize can be devastating to children. For example, losing pets in the wake of a natural disaster can have a particularly potent psychological effect on a child, who saw themselves as responsible for the pet's care (Goldman, personal communication, May 23, 2025).

Goldman also spoke with me about working with Taiwanese children who had seen their schools destroyed and were being taught in temporary trailers (Goldman, personal communication, May 23, 2025). She emphasized the importance of helping these children feel seen and connected with others, such as having younger children create rock gardens and creating community bulletin boards where grade school aged children were able to leave messages to each other. One of the middle-school aged children successfully advocated for having air conditioning in her trailer-classroom. Goldman believes that support groups for children affected by natural disasters could be a meaningful development in mental health services in fostering a spirit of healing through connecting with others.

The human mind can imagine worst case scenarios, an ability that allows us to plan ahead but can also lead to non-productive responses like rumination. Climate-aware mental health professionals have started using the term *pre-traumatic stress* to describe distressing thoughts surrounding catastrophic situations if climate change and environmental devastation continue unabated (Van Susteren, 2021).

Thoughts of anticipated calamity can be reframed into gratitude. As an example, I will share thoughts I recently had when I felt winded during a bike ride to work. Spring had brought an increase in pollen as well as humidity during unseasonably warm days that have become increasingly more common. As I biked up the toughest uphill portion, I struggled to catch my breath. This prompted me to think about how breath is something I take for granted – the constant access to relatively clean air in contrast to the pain of gasping for a breath. I think it is unfathomable that we could lose access to this most basic necessity, but I allow these dark thoughts to co-exist with gratitude for living in an area of the world with relatively less polluted air.

Victims have the potential to join together. It's important that environmental groups take care to embrace people's stories, however varied they might appear to be, and not turn people away simply because they are not sticking to the exact script of that group's agenda. As an example, a story in *The New York Times*, "This is How to Win an Environmental Fight" (Blake, 2025), highlights widespread efforts to push for legislation regulating polyfluoroalkyl substances, or PFAS, also known as forever chemicals. These efforts have united democrats and republicans and people with varying levels of education. Many of these activists were victims who had been sickened by these chemicals or had relatives who were affected adversely by them.

As the mental and physical health effects of a warming earth become more apparent, more victims will be identified. While an increase in victims is a tragedy, there is hope that clearer identification of environmental victims across various contexts will propel the unifying power that collective victimhood can generate.

At the same time, we can't assume that *all* people affected by natural disasters, or any other kind of trauma, are processing their experience through the same mental filter. As with other tragedies, not all victims choose to respond with activism, and this choice needs to be respected.

Even among those who transform their grief into activism, there is no standard prescription for how this looks. Liz Jensen, who founded a literacy activism group one year before her son died unexpectedly in his early 20s, likely from an unknown genetic heart condition, describes how she stepped away from activism initially and later realized that the best way to honor her son would be to carry on the work he had done as a climate activist (Jensen, 2024). She reflects: "Yes, my son was dead. And yes, the planet's life support systems were weakening. But it wasn't too late for the planet" (para.13). However, it's hard to say which direction a proactive response to grief will take. Therefore, we can't assume that environmental tragedies will prompt a unified front of climate change activism.

Contextual aspects framing the extent to which one is personally affected by a natural disaster can affect how strongly they link that event to climate change. A study by Osberghaus and Fugger (2022) showed that people had stronger beliefs in climate change after a flooding event, but only when they lived close to the site of the disaster. Research also shows that people's reactions to natural disasters are shaped by pre-existing beliefs in climate change (e.g. Diakakis et al., 2021). As a case in point, the results of one study (Carlton et al., 2016) showed that victims of drought did not change their perceptions of climate change when the victims didn't believe in climate change before the drought happened.

Other aspects of identity, such as age and gender, might also influence how climate change is considered in the aftermath of natural disasters. For example, McDonald-Harker et al. (2022) discuss cases in which views shifted in a pro-environmental direction in child and adolescent victims of natural disasters. For example, one qualitative study that focused on Filipino youth living in areas that are more prone to flooding showed more problem-solving in response to the hazards of climate change, as opposed to denialism (Berse, 2017). This is consistent with a 2024 survey of US Youth (ages 16- to 25-years-old) showing that self-reported exposure to more types of severe weather events was significantly associated with more climate distress and wanting to take action to fight climate change (Lewandowski et al., 2024).

Regardless, research consistently reveals long-standing mental health effects of natural disasters, including the onset of post-traumatic stress disorder (PTSD), depression, and anxiety and related disorders (Heanoy & Brown, 2024). Treatment for these disorders, whether they emerge after or are exacerbated by being the victim of a natural disaster, is complicated by the diffuse and unpredictable nature of climate change-exacerbated environmental catastrophes. For instance, CBT for PTSD, also known as Trauma-Focused CBT (TF-CBT) entails guiding the client in re-writing their trauma story, reframing unhelpful thoughts such as those of self-blame and helping the individual feel safe (e.g. Ramirez de Arellano et al., 2015). The perpetrator of climate change-exacerbated natural disasters is more diffuse, though, than the whims of a storm or a single perpetrator. It is a situation that we have all contributed to and from which there is no clear escape. This begs the question, does increased awareness of climate change help one recover emotionally from distress caused by natural disasters?

Research sheds light on the role of self-efficacy (or perceived belief in one's ability to manage environmental disasters), hope, and social support in bolstering resilience to natural disasters, at least in the short term (Ime, 2024; Nygaard et al., 2016). Empowering people, then, might represent one critical but alone, insufficient, component for helping people heal from the mental health effects of natural disasters. A climate-change centered approach, then, might be beneficial in that this would identify the root cause of future natural disasters and allow victims to take charge in preventing further suffering.

One theme that emerges here is the idea of proactively coping with the threat of climate change by working to mitigate its effect whether through intellectual or creative projects, activism, or in response to trauma induced by natural disasters. This leads us to our last section on *Strivers*, as we further explore what it means to immerse oneself in climate work, and the challenges that this work presents for those who aim to find a sense of meaning and purpose in it. Seeking salvation through striving comes with a number of caveats that mental health professionals must identify and address in those who are passionate about doing something to fight climate change.

STRIVERS

Persistent striving has the potential to snowball into excessive self-control, with self-control being defined as the ability to inhibit competing urges, impulses, behaviors, or desires and to delay gratification in order to pursue distal goals (Lynch, 2018). Many people are familiar with the problems of too little self-control, such as addiction and impulsive behaviors, while the struggles that come with excessive self-control are underappreciated in our "you're useless unless you're productive" culture. According to RO DBT, too much self-control is associated with social isolation and loneliness, which the former surgeon general Vivek Murthy highlighted as major health threats for Americans (Murthy, 2023).

One feature of excessive self-control is maladaptive perfectionism. Perfectionism expert Thomas Curran (2023) argues that the consumerist, never-good-enough mindset engendered by perfectionism is bad for the planet, as we fill our Amazon carts with products that will make us score higher on the social report card that is graded according to number of material possessions. Striving that is associated with perfectionism can lead to a success-at-all costs mindset, which can make it harder to prioritize pro-environmental behavior. For example, time-saving efforts like frozen dinners and online orders might take priority over pro-environmental ones, such as cooking your own meals.

However, striving can also have an environmental bent. According to the Bureau of Labor Statistics (2025), employment for environmental scientists and specialists is expected to grow by 7% between 2023 and 2033, which is faster than the average rate across all occupations, with wind turbine technicians being among the fastest growing professions in the world. Data has also shown an increase in bachelor's degrees conferred in Environmental Science (Data USA, n.d.). However, an unfortunate side effect is the pressure that Generation Z has expressed in believing they need to save the world. This sentiment was echoed, for example, in *Gen Z Mental Health: Climate Stories*, a short film on Gen Z's emotional responses to the climate crisis (Astronaut Productions, 2022).

In situations in which success is contingent upon finding your place in a supportive community, the striver can find themselves aimlessly wandering. Excessive strivers struggle with isolation, and part of this has to do with an idea that philosopher Jennifer Morton discusses regarding the cost of social mobility, which is often to leave family and friends behind (Vedantam, 2023).

Those who are lucky enough to be accepted to elite universities face their own turmoil with endless competition. In "'We are the Most Rejected Generation'" David Brooks (2025) writes about a Yale student who tried to get into the Existential Threats Initiative which focuses on student discussions about climate change and artificial intelligence, but was rejected for not being an expert on existential threats. How are we to unite in the face of existential crises when merely discussing them requires that people develop an expertise in existentialism?

As mentioned above, I deemed myself to be inept at activism because I'm not seasoned in environmental law and science, and I haven't been successful at getting my friends or colleagues to join environmental groups. The question I have come up against throughout my life is, "I have the heart to help the world, but how do I cultivate the intelligence and kind of personality that attracts people to my cause?"

I reached out to climate scientist and activist Peter Kalmus through the contact form on his website (dated March 31, 2022).

> Dear Dr. Kalmus,
>
> I am a clinical psychologist with 2 young children. I am gravely concerned about climate change and ecological breakdown .... Unfortunately, I come with little knowledge about what I can actually do to stop climate change. Because I am not a scientist, I have read about the science behind climate change in an attempt to educate myself enough to fight for the best causes .... My question is, how do we go about bringing scientists and non-scientists together? Perhaps the better question is, do you believe it is more effective to focus on bringing disciplines together toward common causes, or for people to stick with others from their own discipline, with each discipline working toward the same goals?
>
> Sincerely,
> Jennifer Keluskar

Kalmus promptly responded with the following:

> There are lots of actions that don't have a focus on scientists. The vast majority, I'd say. I have a particular interest in trying to get scientists off the sidelines.

Ultimately, I decided to carve out a niche for environmentalism within my professional endeavors, and Kalmus' message helped me feel more confident in doing so.

Carving my slice of professional pie was fraught with challenges.

I thought my best bet was to provide support for preventing activist burnout, but climate activist agencies have their own mental health support networks, such as Citizens' Climate Lobby's resiliency hub.

Existential concerns, though, have come up in my clinical practice, and this had led to great insights, including enhancing my understanding about how evidence-based approaches such as dialectical behavior therapy (DBT) and Radically Open Dialectical Behavior Therapy (RO DBT) can be incorporated into helping others join in the climate crisis.

Ultimately, I found my place in writing, a solitary activity that doesn't quite involve joining, and yet it does, because I had been attending a Youth subcommittee of the Climate Psychology Alliance of North America (CPA-NA) for a little over a year when I met with fellow member Linda Goldman to discuss my ideas for a book about psychotherapy and climate change.

Goldman is the author of several books about children and trauma, one of which is about climate change angst: *Climate Change and Youth: Turning Grief and Anxiety into Activism* (2022). Linda was supportive of my ambition to write professionally.

We all need people like Linda in our lives who share our vision to find purpose in this mess of a crisis and hamster wheel of a recognition-obsessed world. In addition to Linda's support, I stumbled upon another mentor, Kristina Pikunas, PhD, a clinical psychologist who offers free climate coaching to people like me who are struggling to manifest a project that will allow them to find their purpose in fighting climate change. Kristina's validation and insights helped me keep up the stamina to pursue my dream.

Striving takes risks, such as lost time, energy, and in some cases, friendship.

In moments of self-doubt, I have found comfort in an episode of the podcast *We Can Do Hard Things* in which Oprah Winfrey was the guest (Doyle, 2023). Oprah spoke about the self-help book she co-wrote with Arthur Brooks, *Build the Life You Want: The Art and Science of Getting Happier* (2023) and focused on the theme of detaching from the things we are overly attached to, including success, and needing to be seen as a good person.

Oprah (Doyle, 2023) tells the story of being at her estate in Hawaii during the Lahaina wildfires in 2023. She partnered with Duane "The Rock" Johnson to set up a fund to put money directly in the bank accounts of victims, and she started with a $10 million donation. She did this only to be "hit with this vitriol" about not doling out more money, with people wondering why her house didn't burn down, and even blaming her for the fires (8:51). She was shocked that the public attacked her for trying to do good, although she noted that people on the ground receiving aid were thankful.

Even Oprah, who has been a household name as long as I can remember, feels down when she is misunderstood.

Oprah also shares the experience of making a film about slavery that bombed at the box office. She had wanted millions of viewers to feel the emotions that she felt in response to the story she portrayed. Oprah learned a lesson from this experience:

> And so I learned in that experience that everything else that you ever do in your life is a gift or an offering. You do it with the purest of intention to be a gift and an offering and if people receive it, they receive it. And if they don't receive it, that's OK because your intention was to offer it as a gift. And so that is how I learned to detach from my attachment. (4:51)

Celebrity strivers have achieved fame and money, and even still might struggle to find meaning and purpose in the context of existential threats such as climate change.

Like Oprah, Jeremy Strong is a fascinating case of a celebrity trying to transcend Hollywood and do good for the world, only to receive an unexpected reaction from the public. On March 19, 2024, David Marchese published an exclusive interview with Strong in *The New York Times*. Strong starred in the hit HBO series *Succession*, in which he plays Kendall Roy, an egotistical, success grubbing son of a media mogul, Logan Roy (played by Brian Cox), vying for the chance to succeed him. The family is depicted as the opposite of humanitarian.

In his interview with *The New York Times Magazine* (Marchese, 2024), Strong expressed his long-standing interest in acting in films that explore real social issues. He talked about his upcoming role in a Broadway play called *Enemy of the People* in which a doctor tells his town the truth about an environmental threat and gets criticized instead of praised by the community. Strong reflects on how this role relates to his career, highlighting the tension that arises between wanting to climb up the ladder of popularity and recognition versus wanting to create artistic work that feeds his soul. He says, "I do feel that there's an inverse relationship between creative flourishing and whatever recalibrations you're willing to make to fit into a market …." (para. 6).

A previous interview in *The New Yorker* (Schulman, 2021) had depicted Strong as being arrogant, which was upsetting to him. As an actor, he wanted to channel his emotions into a piece of work, and in Strong's case this meant funneling the betrayal he felt from the interview that labeled him arrogant into the betrayal experienced by the role of the doctor he played in *Enemy of the People* (Marchese, 2024). Strong hoped to heal emotionally as well as repair his public image by engaging in the play.

However, another article published in *The New York Times* just a few days before the interview with Marchese introduced another twist to this story. The headline read: "Climate Protesters Disrupt Broadway Play Starring Jeremy Strong" (Paulson, 2024).

The protestors, who represented a climate activist group known as The Extinction Rebellion, shouted, "'No theater on a dead planet!'" (Paulson, 2024, para. 1). Before security could remove the protestors from the theater, Strong's immediate response was to continue with the show, pretending that the protestors were part of the production. They had some of the audience fooled.

This was a spontaneous joining in the context of tension. There was tension between Jeremy Strong and the protesters, and yet they inadvertently joined Strong by the devices of his creativity. Later in the text we talk about how artists of all kinds have contributed to fighting climate change. The power of creativity can save the earth despite the carbon-gorging art elites coming face to face with environmentalists. We can unite when, just like Strong, we find the loophole for creating symphony out of discordance.

CONCLUSION

Just like no human thinks from one perspective in all contexts, no single category of climate change response fully explains any person in all situations. However, we can highlight examples of how and when these responses move toward or away from cohesion with others.

In the next chapter, we will focus on psychological differences in how people of different ages and generations think and behave, and how these differences are implicated in coming together to address climate change.

REFERENCES

Benanev. (2025, April 21). There's a reason the world's a mess and it's not Trump. *The New York Times*. https://www.nytimes.com/2025/04/21/opinion/trump-tariffs-global-economy.html

Berse, K. (2017). Climate change from the lens of Malolos children: Perception, impact, and adaptation. *Disaster Prevention and Management, 26*(2). http://dx.doi.org/10.1108/DPM-10-2016-0214

Blake, M. (2025). This is how to win an environmental fight. *The New York Times*. https://www.nytimes.com/2025/04/27/opinion/forever-chemicals-pfas-pfoa.html

Britannica Dictionary Online (2025). Activist [definition and meaning]. Obtained on May 18, 2025 from https://www.britannica.com/dictionary/activist

Brooks, A., & Winfrey, O. (2023). *Build the life you want: The art and science of getting happier*. Portfolio.

Brooks, D. (2025, May 15). 'We are the most rejected generation.' *The New York Times*. https://www.nytimes.com/2025/05/15/opinion/rejection-college-youth.html

Bryan, B. T., Thompson, K. N., Goldman-Mellor, S., Moffitt, T., Odgers, C. L., So, S. L. S., Uddin Rahman, M., Wertz, J., Matthews, T., & Arseneault, L. (2024). The socioeconomic consequences of loneliness: Evidence from a nationally representative longitudinal study of young adults. *Social Sciences and Medicine, 345*:116697. https://doi.org/10.1016/j.socscimed.2024.116697

Buckley, C. (2023). After refusing to fly, climate researcher loses his job. *The New York Times.* https://www.nytimes.com/2023/10/12/climate/climate-researcher-no-fly.html

Buckley, C. (2024, February 2). Could a giant parasol in outer space help solve the climate Crisis? *The New York Times.* https://www.nytimes.com/2024/02/02/climate/sun-shade-climate-geoengineering.html

Bureau of Labor Statistics, U.S. Department of Labor, *Occupational Outlook Handbook*, Environmental Scientists and Specialists, at https://www.bls.gov/ooh/life-physical-and-social-science/environmental-scientists-and-specialists.htm (visited May 15, 2025).

Bush, E. (2023). *A scientist reckons with climate grief.* NBC News. https://www.nbcnews.com/science/environment/scientist-reckons-climate-grief-rcna129981

Cadden, C., & Wiens, J. (2017). *The Ongo book: Everyday nonviolence.* Baba Tree.

Carlton, J. S., Mase, A. S., Knutson, C. L., Lemos, M. C., Haigh, T., Todey, D. P., & Prokopy, L. S. (2015). The effects of extreme drought on climate change beliefs, risk perceptions and adaptation attitudes. *Climatic Change, 135*, 211–226. https://doi.org/10.1007/s10584-015-1561-5

Climate Scientist Answers Earth Questions from Twitter (WIRED, October 25, 2022) [video]. *YouTube.* https://www.youtube.com/watch?v=GR46_ohNh9U&t=89s&ab_channel=WIRED

Crooks, J. K. (2024, May 31). *Mindful media: Balancing news exposure with mental health.* West Coast Anxiety Clinic. https://westcoastanxiety.com/mindful-media-balancing-news-exposure-with-mentalhealth/#:~:text=Choose%20Reliable%20Sources:%20Not%20all,to%20manage%20anxiety%20and%20stress

Curran, T. (2023). *The perfection trap: Embracing the power of good enough.* Scribner.

Data USA (n.d.). *Environmental science. Degrees awarded by county.* https://datausa.io/profile/cip/environmental-science#counties_most_degrees

Deb, S. (2024 March 20). Led by its youth, U.S. sinks in world happiness report. *The New York Times.* https://www.nytimes.com/2024/03/20/us/world-happiness-report-finland-us.html

Deb, S. (2025 March 20). Americans are unhappier than ever. Solo dining may be a sign. *The New York Times.* https://www.nytimes.com/2025/03/20/us/americans-solo-dining-happiness.html

Diakakis, M., Skordoulis, M., & Savvidou, E. (2021). The relationships between public risk perceptions of climate change, environmental sensitivity and experience of extreme weather-related disasters: Evidence from Greece. *Water, 13*(2), 2842. https://doi.org/10.3390/w13202842

Domonoske, C. (2018, December 3). David Attenborough warns of "collapse of civilizations" At U.N. Climate Meeting. *Environment.* https://www.npr.org/2018/12/03/672893695/david-attenborough-warns-of-collapse-of-civilizations-at-u-n-climate-meeting

Doyle, G. (Host). (2023, September 21) Oprah shares "the letter from Glennon that freed me" (Season 2, Episode 243) [Audio podcast episode]. In *We can do hard things.* Audacy.

Dreyfus, J. L. (Host). (2023, May 16) Julia gets wise with Gina McCarthy (Season 1, Episode 7) [Audio podcast episode]. In *Wiser than me.* Lemonada.

Dubner, S., & Duckworth, A. (2021, June 13). Why is academic writing so bad? (No. 56) [audio podcast episode]. In *No stupid questions*. Freakonomics Radio Network. https://freakonomics.com/podcast/why-is-academic-writing-so-bad/

Fisher, R. (2020, September 15). The subtle ways that 'clicktivism' shapes the world. *BBC*. http://www.bbc.com/future/article/20200915-the-subtle-ways-that-clicktivism-shapes-the-world

Flavelle, C. (2023, July 16). As climate shocks multiply, designers seek holy grail: Disaster-proof homes. *The New York Times*. https://www.nytimes.com/2023/07/16/climate/climate-geodesic-dome-house.html

Garrett, B. (2021). Doomsday preppers and the architecture of dread. *Geoforum, 127*, 401–411. https://doi.org/10.1016/j.geoforum.2020.03.014

Global Witness (2025). *Land and environmental defenders*. Globalwitness.org/en/campaigns/land-and-environmental-defenders

Goldman, L. (2022). *Climate change and youth: Turning grief and anxiety into activism* (1st ed.). Routledge.

Goldstein-Rose, S. (2020). *The 100% solution: A plan for solving climate change*. HighBridge Audio.

Goulet-Pelletier, J. G., Beaudin, M., & Cousineau, D. (2025). Investigating the negative link between perfectionism and emotional divergent thinking. *Discover Psychology, 5*, 9. https://doi.org/10.1007/s44202-025-00330-x

Gounaridis, D., & Newell, J. P. (2024). The social anatomy of climate change denial in the United States. *Scientific Reports, 14*, 2097. https://doi.org/10.1038/s41598-023-50591-6

Guzmán, M. (2022). *I never thought of it that way: How to have fearlessly curious conversations in dangerously divided times*. BenBella Books.

Hagens, N. (2025, June 3). *Nate Hagens – illusions about the energy transition and our future. Power shift: Shattering illusions about the energy transition and our future [Video]*. YouTube. https://www.youtube.com/watch?v=cT5d8n1Syb8&t=142s

Haltinner, K., & Sarathchandra, D. (2018). Climate change skepticism as a psychological coping strategy. *Sociology Compass, 12* (6). E12686. https://doi.org/10.1111/soc4.12586

Hamann, K. R. S., Wullenkord, M. C., Reese, G., & van Zomeraen, M. (2024). Believing that we can change our world for the better: A triple-A (agent-action-aim) framework of self-efficacy beliefs in the context of collective social and ecological aims. *Personality and Social Psychology Review, 28*(1), 11–53.

Haskett, J. D. (2023). *Solar geoengineering and climate change*. CRS Product (Library of Congress). https://www.congress.gov/crs-product/R47551

Heanoy, E. Z., & Brown, N. R. (2024). Impact of natural disasters on mental health. *Healthcare, 12*, 1812. https://doi.org/10.3390/healthcare12181812

Helliwell, J. F., Layard, R., Sachs, J. D., De Neve, J.-E., Aknin, L. B., & Wang, S. (Eds.). (2025). *World happiness report 2025*. Wellbeing Research Centre, University of Oxford. https://www.worldhappiness.report/ed/2025/

Helliwell, J. F., Layard, R., Sachs, J. D., De Neve, J.-E., Aknin, L. B., & Wang, S. (Eds.). (2026). *World happiness report 2026*. University of Oxford: Wellbeing Research Centre. https://www.worldhappiness.report/ed/2026/

Hoggett, P., & Randall, R. (2018). Engaging with climate change: Comparing the cultures of science and activism. *Environmental Values, 27*(3), 223–243. https://doi.org/10.3197/096327118X15217309300813. (Original work published 2018.)

Ime, Y. (2024). The mediating role of self-efficacy, social support, and hope in the relation between mental health and resilience among 2023 Türkiye earthquake survivors. *Death Studies*, 1–8. https://doi.org/10.1080/07481187.2024.2355240

International Energy Agency (2024). *World energy employment 2024*. www.iea.org
International Energy Agency (2025). *World energy employment 2025*. www.iea.org
Jennett, T. (Director). (2022). *Gen Z mental health: Climate stories! [Film]*. Astronaut Productions.
Jensen, L. (2024, April 8). After I lost my son, I realized I needed to stop looking for closure. *The New York Times*. https://www.nytimes.com/2024/04/08/opinion/grief-five-stages-climate.html
Keluskar, J. (2023, July 7). The psychology of climate change. *Alone Together*. https://www.psychologytoday.com/us/blog/alone-together/202307/the-psychology-of-climate-change
Kennedy, B., & Tyson, A. (2024, December 9). *How Americans view climate change and policies to address the issue*. Pew Research Center. https://www.pewresearch.org/science/2024/12/09/how-americans-view-climate-change-and-policies-to-address-the-issue/
Kenner, R. (Director). (2014). *Merchants of Doubt* [Film]. Participant Media. Madison, A., Cromley, K., & Sharp, M., Executive Producers. (2012–2014). *Doomsday Preppers* (Television Series). National Geographic.
Keshet, M. (2025). Integrative eco art therapy: Addressing humanity's fractured attachment to nature. *Ecopsychology, 17*(2), 131–142. http://dx.doi.org/10.1089/eco.2024.0034
Leiserowitz, A., Maibach, E., Rosenthal, S., & Kotcher, J. (2024). *Climate change in the American mind: Beliefs & attitudes*. Yale Program on Climate Change Communication; George Mason University Center for Climate Change Communication.
Lewandowski, R. E., Clayton, S. D., Olbrich, L., Sakshaug, J. W., Wray, B., Schwartz, S. E. O., Augustinavicius, J., Howe, P. D., Parnes, M., Wright, S., Carpenter, C., Wiśniowski, A., Ruiz, D. P., & Van Susteren, L. (2024). Climate emotions, thoughts, and plans among US adolescents and young adults: A cross-sectional descriptive survey and analysis by political party identification and self-reported exposure to severe weather events. *Lancet Planet Health, 8*, e879–93.
Lynch, T. (2018). *Radically open dialectical behavior therapy: Theory and practice for treating disorders of overcontrol*. Context Press.
Macaluso, C. (2025). Skill remoteness and post-layoff labor market outcomes. *American Economic Journal: Macroeconomics, 17*(2), 134–176. https://doi.org/10.1257/mac.20220062
Marchese, D. (2024, March 19). Jeremy Strong isn't sure he knows who he is. *The New York Times Magazine*. https://www.nytimes.com/interactive/2024/03/10/magazine/jeremy-strong-interview.html?smid=nytcore-ios-share&referringSource=articleShare&sgrp=c-cb
McDonald-Harker, C., Bassi, E. M., & Haney, T. J. (2022). We need to do something about this: Children and youth's post-disaster views on climate change and environmental crisis. *Sociological Inquiry, 92*, 5–33. https://doi.org/10.1111/soin.12381
McKay, A. (Director). (2021). *Don't look up*. Hyperobject Industries.
McKay, M., Davis, M., & Fanning, P. (2021). *Thoughts and feelings: Taking control of your moods and your life*. Harbinger.
Mendoza, J., & Knudson, R. (hosts). (2025, September 26). [Audio episode]. Is Jane Goodall fighting a losing battle? In *The journal*. The Wall Street Journal Channel. https://podcasts.apple.com/us/podcast/is-jane-goodall-fighting-a-losing-battle/id1469394914?i=1000728621479
Merriam-Webster Online Dictionary. *Slacktivism* [Definition and Meaning]. Retrieved May 18, 2025, from https://www.merriam-webster.com/dictionary/

slacktivism#:~:text=slack%C2%B7%E2%80%8Btiv%C2%B7%E2%80%8Bism,real%20steps%20to%20effect%20change

Murthy, V. (2023). *Our epidemic of loneliness and isolation: The U.S surgeon general's advisory on the healing effects of social connection and community.* US Public Health Service.

Murthy, V. [Rich Roll]. (2025, January 16). *Surgeon General's FINAL WARNING: The TOP dangers to Your health [video].* YouTube. https://www.youtube.com/watch?v=A0NyoYUZNAg&t=142s&ab_channel=RichRoll

NRDC (2025). *Gina McCarthy: Former President.* ginamccarthy.com. https://www.nrdc.org/bio/gina-mccarthy

Nygaard, E., Hussaine, A., Siqveland, J., & Heir, T. (2016). General self-efficacy and posttraumatic stress after a natural disaster: A longitudinal study. *BMC Psychology, 4*(15), 1–11 https://doi.org/10.1186/s40359-016-0119-2

Osberghaus, D., & Fugger, C. (2022). Natural disasters and climate change beliefs: The role of distance and prior beliefs. *Global Environmental Change, 74.* https://doi.org/10.1016/j.gloenvcha.2022.102515

Paulson, M. (2024, March 15). Climate protesters disrupt broadway play starring Jeremy Strong. *The New York Times.* https://www.nytimes.com/2024/03/15/theater/climate-protests-broadway-jeremy-strong.html?searchResultPosition=4

Ramirez de Arellano, M. A., Lyman, D. R., Jobe-Shields, L., George, P., Dougherty, R. H., Daniels, A. S., Ghose, S. S., Huang, L., & Delphin-Rittmon, M. E. (2015). Trauma-focused cognitive-behavioral therapy for children and adolescents: Assessing the evidence. *Psychiatric Services, 65,* 5. 561–602. https://doi.org/10.1176/appi.ps.201300255

Reese, R. F., Swank, J. M., & Sturm, D. C. (2023). A national survey of helping professionals on climate change and counseling. *The Journal of Humanistic Counseling, 62*(3), 201–215. https://doi.org/10.1002/johc.12211

Scheffer, M., Borsboom, D., Nieuwenhuis, S., & Westley, F. (2022). Belief traps: Tackling the inertia of harmful beliefs. *Psychological and Cognitive Sciences, 119,* 32. https://doi.org/10.1073/pnas.2203149119

Schulman, M. (2021, December 5). On "Succession," Jeremy Strong doesn't get the joke. *The New Yorker.* https://www.newyorker.com/magazine/2021/12/13/on-succession-jeremy-strong-doesnt-get-the-joke

Schwartz, S. E. O., Benoit, L., Clayton, S., Parnes, M.F., Swenson, L., & Lowe, S. (2022). Climate change anxiety and mental health: Environmental activism as buffer. *Current Psychology, 42,* 16708–16721. https://doi.org/10.1007/s12144-022-02735-6

Seibert, J. (2025 February 5). *4 climate tech trends & innovations in 2025 and beyond.* Climate Insider. https://climateinsider.com/2024/08/26/climate-tech-trends-and-innovations/

Sharot, T., & Sunstein, C. C. (2024, February 25). Why people fail to notice horrors around them. https://www.nytimes.com/2024/02/25/opinion/brain-habituation-horrors.html

Shreve, C., & Fordham, M. (2018). Gender and disasters: Considering children. *Global Alliance for disaster risk reduction & resilience in the education sector (child-centered risk reduction research-into-action brief).* C&A Foundation.

Simon, J. (2024, February 8). *Climate scientist Michael Mann wins defamation case against conservative writers.* NPR. https://www.npr.org/2024/02/08/1230236546/famous-climate-scientist-michael-mann-wins-his-defamation-case

Smith, N., & Thomas, S. J. (2021). Doomsday prepping during the COVID-19 pandemic. *Frontiers in Psychology, 12.* https://doi.org/10.3389/fpsyg.2021.659925

Sparkman, G., Geiger, N., & Weber, E. (2022). Americans experience a false social reality by underestimating popular climate policy support by nearly half. *Nature Communications, 13,* 4779. https://doi.org/10.1038/s41467-022-32412-y

Sturm, D. C., Metz, A., Daniels, J. A., & Stauffer, M. D. (2021). Climate crises: Helping families mitigate, adapt, and transition during disruption (chapter 19). In D. Capuzzi, and M.D. Stauffer (Eds.), *Foundations of couples, marriage and family counseling* (2nd ed.). Wiley. Sustainability Directory (2025, August 21). Participative Efficacy. https://lifestyle.sustainability-directory.com/term/participative-efficacy

Tabuchi, H. (2024, March 1). He wants oil money off campus. She's funded by Exxon. They're friends. *The New York Times.* https://www.nytimes.com/2024/03/01/climate/young-climate-scientists-future.html

Tyson, A., & Kennedy, B. (2023, October 25). *How Americans view future harms from climate change in their community and around the U.S.* Pew Research Center. https://www.pewresearch.org/science/2023/10/25/climate-activism/

United States Centers for Disease Control and Prevention (CDC). (2024, June 18). *Heat and medications – guidance for clinicians.* Retrieved on May 10, 2025 from https://www.cdc.gov/heat-health/hcp/clinical-guidance/heat-and-medications-guidance-for-clinicians.html

Ur-Rehman, & Peltier, E. (2025, August 19). One day of flash floods leaves a devastating death toll in Pakistan. *The New York Times.* https://www.nytimes.com/2025/08/15/world/asia/pakistan-flash-floods-india.html

Vedantam, S. (Host). (2023, June 13). Between two worlds [Audio podcast episode]. In *The hidden brain.* Hidden Brain Media. https://www.hiddenbrain.org/podcast/between-two-worlds/

World Economic Forum. (2025, November 19). *What is the G20 and why does it matter?* https://www.weforum.org/stories/2025/11/g20-summit-what-you-need-to-know/#:~:text=The%20G20%20Summit%20is%20taking%20place%20in,states%20economies'%20commitments%20to%20the%20agreements%20made

Worrall, E. (2023, July 7). *Psychology today: Climate activism will set you free from anxiety.* Watts Up With That? Retrieved May 18, 2025, from https://wattsupwiththat.com/2023/07/07/psychology-today-climate-activism-will-set-you-free-from-anxiety/

Wray, B. (2022). *Generation dread: Finding purpose in an age of climate crisis.* Knopf Canada.

Chapter **2**

How Responses to Climate Change Differ by Generation and Age

Generational clashes are as ancient as human history. As stated in the Netflix series *The Mind Explained*, "For as long as there have been teens, there have been adults bemoaning them" (Dekornfeld, 2021, 0:22–0:27).

Elders' tendency to judge youth is influenced by stereotypes about the cohort of young people at any given time in history. Researchers have referred to the idea of *youngism*, framed by generational bias to describe negative attitudes that older people have toward people in their 20s. For example, one study by Francioli and North (2021) found that older people endorsed colder attitudes toward people who were in their early 20s than toward former generations when they were the same age. While "resourcefulness" was a positive trait they associated with younger people, they assigned "ungrateful" specifically to contemporary young people.

On the other end of the spectrum, consider how elderly people are perceived by younger generations. Through the lens of ageism, western culture has associated old age with sickness, mental and physical frailty, and burden (e.g. Cruikshank, 2013). This affects how the elderly are treated by others as well as their own expectations for what they are capable of.

People of all ages and generations have a unique role in fighting existential threats like climate change. Just as a chess game can't be played well with a board full of pawns or knights alone, we need a diversity of players at the forefront to prevail. This chapter focuses on the tensions that can come up among different generations, and how, in addressing these tensions, a potent synergy can be helpful to both mental health and the planet. Unlike the universal, time-persistent qualities ascribed to certain age groups, generational categories take into account the effect that time, place, and global events have on who a person becomes.

DOI: 10.4324/9781003489665-3

In a report published by Pew Research Center, Dimock (2019) defines generational categories by year of birth, as follows: The Silent Generation (born between 1928 and 1945); Boomers (born between 1946 and 1964), Generation X (born between 1965 and 1980) Millennials (born between 1981 and 1996); and Generation Z (born between 1997 and 2012). Other sources define a younger Generation alpha (e.g. Höfrová & Small, 2024), as those born between 2010 and 2025.

While there are benefits to generational labels, such as inspiring commonality and assessing changing social attitudes and trends, Dimock (2023) also warns against their downsides: (1) Generational categories are not scientifically defined or universally agreed upon. (2) Attempts to place an entire generation, which spans 15–20 years, in a box, lead to stereotyping at the cost of individuality. (3) Discussions tend to focus on differences among generations more than similarities. (4) Stereotypes of generations can be biased toward the experiences of people with higher socioeconomic status SES. (5) Individuals change as they age regardless of which generation they are associated with.

This chapter discusses how climate change affects Generations Z, Millennials (and their Generation Alpha children) and Boomers. I focus on Generation Z, which, as epitomized by activist Greta Thunberg, has been associated with coming to age at the height of public awareness about the climate crisis (e.g. Goldman, 2022). I also focus on Baby Boomers, whom, despite gaining notoriety for climate denialism, in reality include some of the most influential players in climate activism.

In order to deconstruct our assumptions, we need to look at intersecting factors that, research has shown, vary with age. Before I dive into that discussion, though, let's first take a look at what the data say about how different generations are responding to the climate crisis, keeping in mind the caveat described above about overgeneralizing:

WHAT RESEARCH SHOWS ABOUT HOW DIFFERENT GENERATIONS FEEL, THINK, AND ACT IN RESPONSE TO CLIMATE CHANGE

The expert community has unanimously recognized younger generations, Millennial, Gen Z, and the upcoming Generation Alpha as being uniquely affected by climate change in every area of wellbeing, including physical and mental health, quality of life, and economic opportunity (e.g. Goldman, 2022). For example, the results of a descriptive online survey completed by US youth aged 16–25-years-old from all 50 states and Washington, DC revealed that 85% of respondents endorsed being at least moderately worried about climate change and its impacts on people and the planet (Lewandowski et al., 2024). In addition, 57.9% endorsed being very or extremely worried, and 38.3% indicated that feelings about climate change negatively affect their daily lives. The majority of participants endorsed intentions to support aggressive climate policies, indicating that this age group may have a greater sense of agency in fighting climate change.

These results are consistent with a previous, international survey of youth within the same age range of 16–25-years-old (Hickman et al., 2021) which found that 59% of respondents were very or extremely worried about climate change, 84% were at least moderately worried, and more than 45% indicated that their feelings about climate change negatively affected their daily life and functioning. In addition, 75% of respondents endorsed thinking that the future is frightening, and 83% endorsed beliefs that people have failed to take care of the planet. Furthermore, climate anxiety and distress were associated with feelings of betrayal due to inadequate responses from governments.

In the short documentary *Gen Z Mental Health: Climate Stories!* (Jennett, 2022), young adults worldwide are interviewed about their concerns for the planet. Youth express worries about how climate change will impact their future, including their decision to have children, and their access to natural resources imperative for survival.

A mantra of the young climate activists is that older generations have ruined their chance for a livable future with our unsustainable, environmentally destructive lifestyle. Internally, they wrestle with eco-guilt, or the angst that transpires when one reckons with their own inevitable contribution to the same destruction.

In Radically Open Dialectical Behavior Therapy (RO DBT), the principle of *radical openness* (Lynch, 2018b), emphasizes that we see the world as it is, not as we are. What we perceive, what we are told, and what we are led to believe define our conception of reality. Blaming others for the lifestyle they were led to believe was essential for their livelihood is like blaming a fish for swimming in the pond they inhabit. It is like blaming people for smoking in the first half of the 20th century when the tobacco industry launched marketing campaigns aimed at convincing them that smoking was healthy and sophisticated.

Differences in generations are constructed through stereotypes and normalization of labels such as the one espoused by the meme "OK Boomer." It behooves us to be wary of the cognitive bias created by such labels, which lead us to see all behavior of a said group through the filter of that stereotype. Such labels are isolating, fueling the pressures youth feel to fix the climate crisis on their own, which will make their work harder.

Another misconception is that, because younger people will bear the brunt of climate change's effects, including extreme weather events like hurricanes and drought, hazardous heat waves, and limited resources, older people cannot possibly have a stake in the game. But specialized climate activist groups such as Mothers Out Front, The Association of Swiss Senior Women for Climate Protection and Guardians of the Aquifer, the latter two of which largely consist of older women, challenge the view that middle aged and older women don't care.

Coming together means authentically understanding people's differences across age groups, including how people show shifts in certain psychological characteristics over the course of their lives. These factors include perception of time, views on work (including volunteer work), sense of meaning and purpose, emotion regulation skills, and perfectionism.

PERCEPTION OF TIME

In contrast to stereotypes one might conjure of a wrinkled finger waving admonishingly in a younger person's face, an intriguing research study (Cheng & Kim, 2008) highlighted that older people are generally more forgiving than younger people. This may be because of what is known as the *limited time perspective*, or rather, seeing time left in life as limited as opposed to expansive.

Perception of time is relevant to the psychology of climate change. Scientists have informed us that the window of time left to avoid catastrophic effects of global warming is quickly eclipsing (IPCC 2023), and we need to act right away to avoid its worst effects.

How one reacts to the urgency conveyed by experts may in part depend on a concept known as *time horizon*, or how limited versus expansive you perceive your time left on earth (Carstensen, 2021). Research by psychologist Laura Carstensen looks at the effect of shifts in one's time horizon, from seeing time as expansive when people are younger, to seeing time as limited as they age.

This research (Carstensen & Reynolds, 2023) has shown that a limited time perspective is related to positive emotional outcomes, such as less worry, and that this is due to a concept known as *socioemotional selectivity theory* or SST, in which finding pleasure in the present is prioritized over sacrifice in the hopes of reaping future rewards. Thus, older people seek out experiences that will give them more pleasure in the moment over making decisions based on anticipated future payoffs.

SST (Carstensen & Reynolds, 2023) explains why older people prioritize close friendships over new ones. It turns out that younger people shift to prioritizing close friendships over forming new relationships just like people of more advanced age when they are asked to imagine a scenario that implies time constraint, such as imagining they will soon move out of state. On the other hand, when older people are put in conditions of time expansion, such as being asked to imagine that a new medical treatment will extend their life, they shift to showing preferences that are more typical of younger people. Therefore, Carstensen's research shows that we can shift to a different time horizon simply by changing the way we think, rather than waiting until old age to adopt it.

Elders can inspire younger environmentalists by emphasizing the power of living in the present. Jane Goodall's grandson, Merlin van Lawick, who worked with Goodall in conservation, notes, "I like to focus on living for the moment and doing my best. I learned that from my grandmother, and it resonated with me" (Hirsh, 2023, para. 19). Lawick describes how trying to solve everything all at once can fuel apathy. In contrast, by completing smaller tasks that are also in line with what you believe in, such micro-accomplishments can build up over time into a legacy of making a difference.

Hope in having a future is a typical characteristic of the young and healthy, even if one is pessimistic that it will likely be a destructive future. The trouble faced by younger adults, rather, tends to lie in fear of uncertainty, such as uncertainty in the sustainability of their employment and in being able to save enough money. My young adult clients tend to focus exhaustively on trying to complete their career goals and save as much money as possible before social security runs out and the world becomes less inhabitable. Inevitably they feel frustrated in their progress, as such objectives take decades for most people to accomplish.

In contrast, limited time perspective pulls people to the present when their futures, whether anticipated to be good or bad, are dwindling all the same. In Chapter 1, I talked about Gina McCarthy, the first National Climate Advisor from 2021 to 2022, who was interviewed by Julia Louis Dreyfus on her podcast *Wiser than me* (Dreyfus, 2023). Each episode procures wisdom from elderly women with powerful positions.

McCarthy (Dreyfus, 2023) urges young women to follow her mother's advice of not wasting a good worry on things that are outside of their control. In a time limited perspective, control is relinquished and exchanged for a spirit of acceptance as opposed to hopelessness. In fact, McCarthy is described as a professional who keeps on fighting in spite of the bleak prospects that climate change presents. As she does with all her guests, Dreyfus asks McCarthy how old she *feels*, to which McCarthy says she feels 32, when she is actually 71-years-old at the time of the interview.

McCarthy (Dreyfus, 2023) says she wants younger women to speak up in their workplaces, and if they are not heard the first time, they should speak up again. At the same time, she emphasizes the power of slowing down and taking one step at a time. When you take small steps, bigger ones follow. McCarthy discusses how people tend to become fixated on needing to know whether their endeavors will be successful, and this slows them down from getting started. She contends that the only thing you need to know is how to *get started* and not how you will finish.

Needing to know how things will turn out in order to start an endeavor speaks to another aspect of time perception called *time poverty*, the idea of there being a deficit of free time in many working people's lives (Harvey & Mukhopadhyay, 2007), but it can also refer to a belief, to whatever extent self-imposed, that one never has enough time to do everything they want or need to do. Someone who believes themselves to be time poor might be more vulnerable to stalling on endeavors when their outcome is uncertain. In other words, if it's not clear how the long, hard endeavor will work out in the end, you risk wasting a lot of time by pursuing it.

According to the opportunity cost model (Kurzban et al., 2013), each moment spent on a given activity is a moment taken away from another way you could be spending your time. When we don't feel we are getting enough out of an endeavor to warrant the time spent on it, therefore, we abandon the task. This speaks to a topic I wrote about in a blog post for *Psychology Today* about "All-for-Nothing Angst," which I refer to as AFNA in the post (Keluskar, 2024). I provide the following tips for coping with the dread of spending time you didn't think you had to begin with on a task that did not pan out as planned:

1. *Accept that AFNA is part of life.* Everyone has had all-for-nothing moments.
2. *Have hope in recovering mental energy.* While you might not be able to avoid the pinch of AFNA, you can foster hope that your motivation will come again. In the meanwhile, be willing to file your efforts (mentally or by writing down the experience and saving it in an actual file) for potential later use.
3. *Link your angst to a need.* If wasted effort is causing you more angst than you think it should, it could signal that you have an unmet need, such as to rest, slow down, or seek encouragement from others.
4. *Budget your mental energy like you manage time.* Monitor your mental energy expenditure. Taking time to recharge and spreading out mentally taxing work will help prevent all-for-nothing moments and put them in perspective when they happen.

Believing we don't have enough time gets in the way of living in accordance with values, including pro-environmental ones like writing to a representative about promoting a bill for clean energy.

According to the concept of the value-action gap (Franco & Ghisetti, 2022), people don't always act in a way that reflects their values, such as running water too long when you are concerned about conserving it.

People might believe that environmentalism and addressing climate change are critical objectives and at the same time fail to act on them due to a perception of lacking time wealth.

The leader of a climate activism group I was once a part of bemoaned that people need to be able to afford not having a paying job in order to be able to commit sufficiently to the climate cause. When one perceives themselves as lacking in discretionary time, or in other words having less time wealth, they are less likely to volunteer time toward a cause.

In relation to the climate crisis, the person who sees themselves as lacking discretionary time is less likely to invest the time to go against a system that values productivity at the expense of the environment. For instance, choosing to use cloth or disposable napkins means an extra load of laundry without knowing for sure that it is going to have a meaningful effect on helping the environment.

Discretionary time and time wealth, to an extent, overlap with monetary wealth, and not every elderly person has sufficient amounts of the latter saved up to afford to retire (e.g. Cruikshank, 2013). The idea of a time-wealth disparity was poignantly highlighted in a piece by Esau McCauley, an opinion writer for *The New York Times*, who talked about his mother having to work long hours to make ends meet (McCauley, 2023). He and his siblings were watched by friends and family, highlighting that, while the time poverty of the materially poor is a social problem resulting from not providing enough community supports and pay for laborers, on the other side of the coin, the idea of time poverty pushing individuals to rely more on community could be seen as a sign of resilience.

Ultimately, control over your time and efforts enhances both mental and physical health, mindfulness is a skill that is free to all and that can help you stay in the present. It involves observing your own thoughts and identifying when the thoughts entail judgment, such as critical words over how you are spending your time or the environmental cost of saving time by running an extra load of laundry. We can observe thoughts and the feelings that accompany them, such as anxiety, guilt and longing, etc. We can approach these thoughts and emotions with curiosity rather than judgment. While mindfulness in itself is not the same as problem-solving, it can pave the way for proactively addressing problems.

You can ask yourself, how can I compensate for not having enough time to [engage in this sustainable behavior]? Could I lean into my community for support? Could I create routines and habits that save time? One of my favorite conservation efforts, although an infrequent one due to not eating out a lot, is bringing my own non-disposable Tupperware to restaurants instead of using their takeaway containers. When I forget to bring them, I re-use takeaway containers for small household items, like unused squares of parchment paper that could be used later.

I've also incorporated small, easy habits into my daily routine such as shutting off the faucet while I'm scrubbing pots, tearing off a small square of paper towel instead of using the entire thing, and making it a routine to ask myself, "Is there another way I could use this item before I trash or launder it?"

These acts constitute *micro-sustainability* (Imolore, 2023). When I first became interested in environmentalism about a decade ago, people thought such acts were ludicrous because of their micro-effects. The way to boost morale, though, is to enhance the joy we can derive from them. I liken this joy to one that builds with time and commitment, similar to how people might need to force themselves to work out initially but then come to crave exercise.

Elders can help foster this mentality. My grandparents, for instance, grew up in Sicily during World War II. They didn't know about climate change, but they developed a spirit of conservation due to limited resources. Even after they acquired a better quality of life in the United States, they carried on this spirit in subtle ways, like re-using jam jars for saving coffee in the fridge.

Frank Pinello, owner of the restaurant Best Pizza in Williamsburg, Brooklyn, features a video on *Munchies,* a *YouTube* food channel (2015) in which his Nonna, who was from the same small Sicilian town as my grandparents, makes chicken cutlets and re-uses a bottle of maple syrup for olive oil, which is functional in preventing too much oil from coming out at once. We can observe and honor these efforts as pro-environmental regardless of whether their intention was to save the planet or to save a few bucks on olive oil.

Older people who are able to retire might have more discretionary time to devote to volunteering and advocacy. They might also be willing to trade ambition for relationships. In this sense, older activists have much to offer their younger counterparts and can serve as valuable mentors.

Elders can also offer a more measured approach to advocacy that helps prevent burnout, a subject I explore further next within the topic of work.

WORK

Work settings represent employees who might range in age by decades. There is potential for older employees to contribute the wisdom of their lived experience in a profession, and for younger individuals to bring fresh ideas to the table.

Solomon Goldstein-Rose has been a climate activist since he was 11-years-old, and he served on the Massachusetts state legislature at the age of 23 (https://www.solomongr.com/). In his book, *The 100% Solution: A Plan for Solving Climate Change*, Goldstein-Rose describes how, when examining a problem, older people are more likely than younger people to think about it within the context of their life experience.

Intergenerational tensions can still arise, though, and we need tools to work through them. For example, when I talk to elder adults about my eco-distress, they sometimes reference existential crises they grew up during, such as the Cold War, as evidence that these things generally blow over. Others have leaned into the power of technology to solve world problems. This perspective is understandable given the tremendous advancements since the mid-20th century that they have lived through.

While meant to be reassuring, these sentiments might feel invalidating to those who see the unique threats of climate change, and especially to youth, who will live to see the worst effects of the climate crisis.

However, Goldstein-Rose (2020) points out that while young people have yet to acquire the perspective of life experience that elders possess, younger generations' forward thinking can be an asset. He writes, "Political rhetoric may return repeatedly to the same ideas, but technology keeps moving in one direction" (p. 16). Selling the technological advances that are consistent with a greener earth, then, could be one avenue for connection between younger adults and elders.

Goldstein-Rose's (2020) point about the potential of young adults to be the forebearers of cutting-edge technology bears resemblance to anthropologist Margaret Mead's concepts of prefigurative, post-figurative, cofigurative cultures (APA Dictionary of Psychology, 2018a–c). In prefigurative cultures, people typically learn from those younger than themselves, as evidenced by elders asking younger people for help with technology. In contrast, postfigurative cultures entail the young learning chiefly from their parents, grandparents, and other adults. This represented society for most of human history. In cofigurative cultures, people learn mostly from other people in the same age group. Modern society demands that we find a blend of all three. Technology is moving at a rapid pace and being mastered at younger ages. By the time she was nine-years-old, my daughter was quickly surpassing my abilities to navigate apps and computer systems.

In addition, as noted by Linda Goldman (2022) in her book *Climate Change and Youth: Turning Grief and Anxiety into Activism*, younger generations may be more literate on scientific data related to global warming and can become frustrated when change does not happen more quickly. Elders can serve as mentors in honing writing and oral communication skills to help younger people realize the power of getting the word out in language that engages their audience.

While there is no denying that we will benefit from the tutorship of younger people when it comes to technology and knowledge about environmental crises, they will also continue to need elders for wisdom, which comes from experiential learning. However, experiential wisdom is not a concrete, marketable, profit-making good. This makes it harder for employers to support mentorship programs that do not add up to immediate economic growth in their sector.

But depriving younger people of mentorship might contribute to burnout. A 2022 study revealed that a whopping 98% of Gen Z employees indicated symptoms of work burnout (Cigna Health, 2022). While this is largely due to financial pressures, uncertainty about the future is another driving force that is probably, in turn, exacerbating financial pressures.

The pressure to save money for a dire future takes up much mental space for my young adult clients, who also tend to express the belief that they must accrue wealth first before engaging in humanitarian endeavors, including helping the climate crisis. These pressures are sucking the meaning and enjoyment out of their work experience. This is consistent with the findings from the study above (Cigna Health, 2022) which also showed that Gen Z, compared to Millennials and Boomers, scored the lowest on an assessment of work satisfaction known as SCARF, which stands for Status (feeling valued at work), Certainty (knowing where they stand as employees), Autonomy (having choices over how one works and being trusted to make decisions), Relatedness (sense of belonging and fitting in at work) and Fairness (being treated with respect and fairness).

I encourage clients struggling with low SCARF to seek out creative passions, even if this is outside of the office. A few years ago creating a website of mental health resources on my own was ultimately what saved me from burning out. At the time, I didn't know for sure that this would be a panacea for my emotional burnout, but I felt compelled to try. I had to lean into hope, combined with seeking mentors who helped me balance my passion with logistical measures, such as getting feedback on the quality of my writing and the layout of my website.

Jane Goodall, the iconic elder environmentalist, offered a perspective of how hope can be cultivated even when the odds seem stacked against us. In the book she co-wrote with Douglas Abrams: *The Book of Hope: A Survival Guide for Trying Times* (2021), Goodall reflects on global atrocities that transpired during her lifetime, such as the Holocaust. She says that despite the enormity of today's problems, evidence abounds that many still care about social justice.

Goodall and Abrams (2021) define the concept of hope as "What enables us to keep going in the face of adversity. It's what we desire to happen, but we must be prepared to work hard to make it so …" (Part I, 18:07–18:18). She explains that hope does not in and of itself require action. A prisoner hoping to be released, for instance, is rooted in the desire for others to take action. Animals can hope. Chimps, Goodall explains, can throw a tantrum when they don't get what they want, which could be considered a frustrated kind of hope. In humans, hope can also serve as a survival trait, without which we would perish. It can generate action, and action can sustain hope.

We return to the idea of hope and the combining of hope and agency, referred to as active hope, in Chapter 4. For now, let's explore whether the bidirectional relationship between action and hope that Goodall describes can generate synergy between different age groups that share a concern for human survival. And could elders be the beacons of hope, deepening our understanding of this construct?

Abrams (Goodall & Abrams, 2021) tells the story of Goodall's grandson, Merlin, who at the age of 18 dove into an empty swimming pool and suffered severe injuries. This event motivated him to give up partying, and, along with his sister, Angel, join their grandmother in conservation work. This story is an apt transition to what I want to talk about next, meaning and purpose, and how intergenerational work can help us fulfill these human needs.

MEANING AND PURPOSE

Differences in how youth and elders approach activism illustrate the potential of elders to impart tools for compassion that could help advance a cause. Whereas youth lean into a condemning tone in their entreaties to government officials, older people tend to lead with love, such as pushing for change with gentle acts of kindness (Pipher, 2023). Unlike their younger counterparts, who seek meaning and purpose through recognition and fitting in, elders seek these psychological needs in more holistic ways. As Dr. Mary Pipher puts it in her *New York Times* piece *Grandmothers of the World, Unite*:

> Indigenous people's rights, social justice, local control of food and energy, and democracy are all part of one great cause – the respect for and preservation of all life. We are the mother trees in the forest, nurturing all our surroundings. (para. 11)

While elders, particularly women who might have had different career paths due to taking the time to raise children, stand to impart to youth a more holistic perspective of life success, they also might find greater meaning and purpose in fighting to preserve the resources they wish to pass down to their brethren. In this sense, All for Nothing Angst AFNA motivates efforts to salvage one's legacy. The ideas of legacy and leading with kindness are included in the main message of the platforms for elder climate activist groups such as Elders Climate Action (https://www.eldersclimateaction.org/whyelders) and 1000 Grandmothers (https://www.1000grandmothers.com/).

While young activists rightfully decry their generation being the most affected by climate change, Pipher (2023) describes how elders have a stake in the climate crisis as well. Elders risk greater mortality from hotter temperatures (CDC, 2024). Elders also risk being more socioeconomically disadvantaged (e.g. Cruikshank, 2013), which is why younger people are stressed about saving enough for retirement and hearing news about Social Security's impending extinction. Financial stress will make it even more difficult to obtain limited resources, let alone enough wealth for air conditioning. Unlivable conditions will increase the risk of social isolation, which already stands to shorten lifespans (Murthy, 2023).

Despite these hazards, as Pipher (2023) puts it, elders are keener on leading with compassion than with anger. Pipher, who is part of an environmental activism group called Guardians of the Aquifer, has seen the power of bringing baked goods to legislators who will drive the changes we need for a more environmentally sustainable future. In so doing, elder women's activism is authentic in portraying their role as mothers of the earth.

Rather than seeing an older person's more positive or hopeful approach to the climate crisis as a perspective they can afford to have, younger people can entertain the possibility that the approach of elders represents the product of psychological changes that lead to developing a more holistic perspective.

At the same time, elders might occasionally need to feel the heat of their younger counter parts emotions to drive action. Linda Goldman, a Baby Boomer, built a career on writing about how to help children with grief. However, her son-in-law, who is a Millennial, encouraged her to write a book about how to help children cope with climate change (Communication with L. Goldman, May 23, 2025), a topic that a grief expert such as Goldman is more than qualified to dive into.

Using the full range of emotional intensity to get us closer to our goals demands that we respond to emotions in accepting and yet strategic ways, hence why it is important to look at age differences in emotion regulation.

EMOTION REGULATION AND AGE

In general, people enjoy better emotional well-being in their older adult years compared to their younger ones (Carstensen et al., 2011). Laura Carstensen and her colleagues examined the question of how the ability to manage emotions, such as self-soothing when upset and maintaining positivity in the face of adversity, changes as we age. They used a gold standard approach known as the experience-sampling method in which participants are prompted with pagers to record data. In this case, participants were prompted to report on the extent to which they were experiencing several emotions on a 1-to-7-point sale, including 8 positive emotions, such as happiness, excitement, and amusement, and 11 negative ones, such as anger, sadness, and boredom. The researchers also controlled for other factors that could affect emotional well-being, including personality, verbal fluency (naming as many animals as you could within 90 seconds), physical health, and demographic variables. The results showed not only that older adults have more positive experiences than younger adults, but that their emotions are also more *stable*, or less volatile. Of note is that the sampling occurred during a typical week and did not consider reactions to major life events.

The researchers concluded that older adults are less reactive to the daily ups and downs of life, and they surmised that the mechanism for this is through increased incidence of experiencing positive and negative emotions concurrently (Carstensen et al., 2011). An example might be missing a friend who has moved away but also relishing in the positive experiences you have shared with them. Interestingly, the intensity of emotional experiences did not change with age, so emotional suppression was not at the root of these findings.

As noted in the above section on time perception, shifts in time horizon that occur with age are associated with increases in positive affect (Carstensen et al., 2011). Carstensen ties the same trend to improvements in emotional stability. As people age, they see time on earth as more limited and therefore focus more on maximizing positive emotions. However, the behavioral changes that result from shifting time horizons are important to consider as well, such as the increased importance placed on close relationships, which could relieve isolation and loneliness. Relationships can help us with emotion regulation because friends provide emotional support, such as validation, as well as information, such as ideas for solutions to everyday hassles.

In addition to going hand in hand with enhanced well-being, emotion regulation skills can contribute to better work performance. Former justice Ruth Bader Ginsberg owed much of her successful and lengthy tenure to her emotion regulation skills. She saw anger as a useless emotion that dilutes one's power to persuade. When she felt angry, she found ways to diffuse it such as playing the piano (Ross, 2020). Similar to McCarthy's bridge-building approach, she also found purpose in helping people.

Thinking about helping others in a holistic way furthered Ginsberg's feminist agenda in unexpected ways. In one famous case in the 1970s (Romero, 2020), Ginsburg defended a man named Stephen Wiesenfeld whose deceased wife had been the family's primary earner. Wiesenfeld was told he didn't qualify for survivors' benefits because he was a man. Ginsburg won the case by arguing that the decision was sexist and therefore unconstitutional. In defending a man, she advanced the broader agenda of not denying people privileges based on their sex or gender.

Holistic thinking goes hand in hand with improved emotion regulation, and this is an essential area where cognitive behavioral interventions can help people get past emotional hurdles that are limiting their effectiveness to engage in the climate crisis.

In her interview with Julia Louis Dreyfus (2023), Gina McCarthy gives advice to young women that echoes the importance of balancing flexibility with persistence in advancing a broader agenda. She emphasizes, "So you've just got to go with the flow, but also recognize that there's a ground you need to keep as a human being. There's a respect you need to demand, especially in political situations" (12:09–12:22). McCarthy emphasizes that she doesn't get emotional at work, but she has reacted to disrespect such as sternly telling a colleague not to refer to her as "Sweetheart." When asked what the best career advice she's ever been given was, she reflected on a time the then Chief of Staff of Environmental Affairs said that you should never push the question if the answer is going to be "No:"

> Keep plugging. Because it will break. But if you try too soon. If you push too hard. And because you're frustrated, not because you've found a way to argue something different, then you're going to lose. So every time from then, when I hit a wall, I've thought to myself, 'What's another way to get it? What do I do different that's going to start a separate conversation that will get me where I need to go?'
>
> (Dreyfus, 2023, 16:10–16:42)

McCarthy's mindset embodies the RO DBT concept of Flexible Mind, which emphasizes shifting course. This is neither giving up or digging your heels into self- righteousness, as an ideal path to pursue in the face of adversity (Lynch, 2018b). This concept comes up often in my discussions with adolescents and young adults with existential concerns. They vacillate between, on one hand, feeling frustrated with the world for not caring about the urgency of these concerns and shame at not being "good enough" either to solve world problems or to get ahead of them, such as through saving money or prepping behavior (discussed in Chapter 1). McCarthy, in line with the elder mindset I've described in this chapter, urges people to take one step at a time and emphasizes that you have to start somewhere, even if you don't know for sure where your efforts will lead you.

RO DBT skills, such as radical openness, which prompts clients to consider how another person is perceiving a situation and understand their reactions (Lynch, 2018b), can be applied to addressing generational differences in the response to the climate crisis. Radical openness allows people to hold their ground respectfully and tactfully by acknowledging others' perspectives even if they don't completely agree with them.

McCarthy conveys radical openness when she talks about insensitive politicians and their bluster:

> You do what you can to be as respectful as you can. But you don't ever have to agree [with them]. And you don't ever have to try to bounce back and be as nasty as someone was to you. It's the worst thing in world, especially for a woman. That's not the atmosphere within which you can win. (19:30–19:52)

Dreyfus (2023) and McCarthy acknowledge the unfairness shown toward women in politics, but McCarthy shows that you can tactfully work with these realities and still get to where you need to go.

Given that RO DBT is an approach that specifically targets maladaptive perfectionism (Lynch, 2018b), the usefulness of these skills is further highlighted when we look at how perfectionism interferes with environmental values. The next section explores this, and how collaborating with different generations and people from other walks of life in general can help us overcome the dark side of perfectionism.

WHEN PERFECTIONISM INTERFERES WITH ENVIRONMENTAL VALUES

Perfectionism and environmentalism, two aspects of my identity, are in a tumultuous sibling relationship, bickering one moment and united, but ganging up on their parents, the next.

My tendency to strive for perfection gets in the way of my environmental values in several ways. For one, the perfectionistic tendency to try to optimize time, that is to do as much as possible in as little time as possible, conflicts with pro-environmental behavior, such as choosing take-out over home-cooked meals. Perfectionism can also manifest in how we go about expressing environmental values, such as timing household showers. If overblown, this pairing can backfire. Being perfectionistic in my pro-environmental behavior risks turning people off, such as when my kids think they are in trouble for being told their showers are too long, which sabotages the collective action needed for curbing climate change.

In a conversation with my friend and colleague Linda Goldman (personal communication, May 23, 2025), the author of *Climate Change and Youth: Turning Grief and Anxiety into Activism* (2022), I asked her for advice about how to get my kids to engage in more sustainable behaviors, like taking shorter showers. I wanted to get them to behave more sustainably without coming off as restrictive or controlling, or in other words, without letting perfectionism get in the way. The last thing I wanted was for them to think that me telling them to stop wasting water was the same as me saying they weren't good enough.

Goldman emphasized the importance of language when interacting with children (personal communication, May 23, 2025), including when it comes to discouraging perfectionism. Goldman sees perfectionism as a grief and loss issue because perfectionism "doesn't give them that space, that breath to enjoy the nature of who they are ... Nature is perfect in its imperfection, and so are children."

She advised using language such as, "We are working as a team to help the planet. Do you have any ideas?" In addition to advising giving children control over coming up with ideas, she suggested leaning into experiencing nature together with children, such as growing vegetables and letting kids see how these vegetables eventually make it to the dinner table. Fostering a connection with nature inspires a conservationist spirit for children without them having to experience the "heavy burden of never being good enough" (Goldman, personal communication, May 23, 2025) that inevitably comes with perfectionism.

The benefits that taking the edge off maladaptive perfectionism has for the planet run both ways. If we live by environmental values in a way that is flexible and focused on strengthening social connections, there is potential for pro-environmental behavior to help us overcome the perils of perfectionism, such as burnout and mental health problems, as well as for a more flexible approach allowing us to fulfill our potential for addressing environmental problems.

My aim in this section is to speak to both individuals who are struggling with maladaptive perfectionism and might benefit from shifting their values in a pro-environmental direction, as well as environmental advocates who are experiencing what some have called eco-perfectionism, or over-striving to be flawlessly sustainable.

In addition, I wish to tie perfectionism to age and generational differences, which is why I discuss it here as well as in Chapter 6, which focuses more on RO DBT strategies for addressing eco-perfectionism.

Research using community (i.e. non-clinical) samples has shown that younger adults endorse more perfectionism than older adults. For example, in a study by Robinson et al. (2021), both undergraduates and older adult participants completed measures of perfectionism, conscientiousness (e.g. the trait that allows us to stay on top of our chores and tend to our health), and neuroticism (tending to be anxious and have a negative mood). The results revealed that undergraduates were higher on perfectionistic strivings and neuroticism than older adults, while older adults were higher on conscientiousness, which is an adaptive and helpful trait.

In this case, age differences might overlap with generational differences. Research (Curran & Hill, 2019) shows that perfectionism in college students has steadily increased since 1989. Furthermore, perfectionism expert Thomas Curran, who is a co-author of the analysis on college students' perfectionism, and also the author of the 2023 book *The Perfection Trap*, provides insights that reveal striking parallels between perfectionism and the climate crisis. For one, Curran describes the relationship between perfectionism and consumerism, which has undoubtedly contributed to global warming. He also recommends spending more time in nature to moderate exposure to social media, which, as he discusses, exacerbates the negative impact of perfectionism.

According to Curran, perfectionism is fueled by insecurity, specifically the belief that one will never be good enough. In this vein, people become obsessed with buying products that are marketed to appeal to the part of us that feels we are not attractive, smart, productive, or [fill in the blanks] enough.

Research (e.g. Franco & Ghisetti, 2022) also has shown that those who hold themselves to unrealistic standards in productivity are more likely to perceive themselves as never having enough time. This applies to environmentalism because it might mean that, unless your work is directly related to sustainability or climate activism, perfectionism takes away from the time one needs to engage in voluntary advocacy efforts, such as participating in an action call to a government representative. My clients' conviction to earn a lot of money *before* engaging in efforts to address climate change is an example of how perfectionism holds us back from soothing the ache of urgency this crisis presents.

Perfectionism also fuels decision paralysis. For instance, overthinking might lead one to procrastinate on installing solar paneling in their home. The thought, *What if the neighbors don't like it?* is an example of what Curran and Hill (2019) refer to as socially prescribed perfectionism or an excessive concern with gaining others' approval.

Perfectionism is also associated with trouble taking healthy risks (Lynch, 2018a) due to fear of uncertainty, which might manifest in worrying about transitioning to an electric car when one has driven only gas cars. Perfectionists might also be prone to fatalistic thinking under ambiguous circumstances, such as, "What's the use of trying if others aren't going to follow suit?" The need to please others can exacerbate this kind of thinking as well given that in general, people aren't given enough positive recognition for pro-environmental behavior.

Energy efficiency is also sacrificed in the whirlwind of perfectionism. Perfectionists tend to take the longer route, literally, to the frustration of many friends in the past who begged me to try a short cut when I drove them somewhere. I refused out of fear of getting lost, or in other words, looking imperfect.

Perfectionists will do just about anything to avoid mistakes and criticism from others. This could get in the way of environmental values, such as throwing out recyclable items to make sure there isn't any clutter in the house or being reluctant to join a climate activism group because you aren't up on environmental law and science.

Eco-perfectionism is a parallel but distinct phenomenon. While it is not an actual diagnosis, the concept is validating to people like me, who have brought their baked-in perfectionistic tendencies into sustainable intentions. Ironically, though, this mindset can lead to inadvertently sabotaging sustainability efforts. The reason why eco-perfectionism limits progress is that it interferes with being able to join with others, and this is bad news for the climate movement, which thrives on collective action. It is less likely that a family will collectively commit to pro-environmental behaviors if they continue to squabble over who forgot to recycle.

I have quit several climate advocacy projects because I didn't think I met my peers' expectations. My problem here is in line with socially prescribed perfectionism. In turn, activists who show what is known as other-oriented perfectionism, or having unrealistic expectations of others (Curran & Hill, 2019), might sour their relationship with followers.

Other-oriented perfectionism can be directed at relatives in demanding that their environmental views align with yours. Individuals who strongly value family but are frustrated by their lack of acknowledgement of climate change, thus, become internally conflicted.

At the same time, research (Lawson et al., 2019) has shown that child to parent intergenerational learning, or the transfer of knowledge, attitudes or behaviors from children to parents may help parents express higher levels of climate change concern. Therefore, addressing maladaptive perfectionism can be one way of maximizing the effect of intergenerational collaboration.

Luckily, there are research-based interventions for helping people soothe excessive perfectionism, such as RO DBT (Lynch, 2018a). Although the effect these strategies have on strengthening climate action has yet to be researched, hypothetically, there is reason to believe they can be applied.

The following case study represents a conglomeration of difficulties that young adult clients have brought up. The name of the client below and other identifying information have been made up:

CASE STUDY

Ernie is a young adult living with his parents. He chose a local community college in order to save money, with the plan of transferring to a four-year university with a strong Premedical program. However, living with his parents became more oppressive when they didn't accept his views on environmentalism. They refused his suggestions that they compost their food and use less electricity, and then accused him of being ungrateful that, because of their hard work, they were able to provide him with commodities that they could not afford during their youth.

When Ernie attended a climate protest, his parents balked at the idea and tried to convince him not to go. This green energy nonsense would surely hurt working-class people who had fought for a roof over their heads.

Ernie described feelings of distress over his parents' beliefs, which flew in the face of the humanitarian and environmental values he espoused. How could they be so ignorant? He also felt conflicted because he knew how much they had struggled, and he aimed to please his parents, who had sacrificed an easy life for his own good. What if they got angry at him for disagreeing?

Ernie's feelings of frustration, sadness and guilt were validated, and he was praised for trying to discuss his views with his parents. At the same time, he was guided in understanding the pros and cons of assigning blame (a defensive reaction), or blaming himself (ineffective self-punishment), versus, alternatively, validating his emotions while relinquishing control over things he cannot control: his parents' beliefs when they didn't respond well to the information he shared with them about the climate crisis.

Next, Ernie was guided through Radical Openness DEF skills (Lynch, 2018b). He was asked to accept negative feelings instead of trying to get rid of them (which only makes them more intense), and to acknowledge and describe his emotions of hurt, frustration, and anxiety about being rejected by his parents (D for describe emotions), even if in a journal.

E is for self-enquiry, a process of self-reflection that entails asking yourself questions that challenge your assumptions and scanning your emotions for a response to these questions that resembles your "edge," a feeling of unease that indicates an unsavory truth about yourself, such as demanding that others change instead of focusing on your own actions. Clients are instructed not to try to rush to answers or solutions, or to soothe the discomfort. Rather, they are encouraged to generate questions to come back to and reflect on.

Clients practice self-enquiry (SE) on their own, but a therapist can offer suggestions for initial prompts/questions with which to start the process. Ernie decided to consider the following SE questions (from the RO DBT treatment manual, Lynch, 2018b). Note that the term "disconfirming feedback" refers to any viewpoint that contrasts with what you believe to be reality, including things we know to be facts, such as the realities of climate change.

> Do I believe that I know what the intentions are of the person giving me the disconfirming feedback? For example, am I assuming that they are trying to promote themselves? Or do I believe that they are trying to manipulate, coerce, or intimidate me? If yes or maybe, then is it possible that I am not really giving them a chance? What am I afraid might happen if I were to momentarily drop my perspective?
>
> (Lynch, 2018b, p. 63)

Clients are urged not to rush to judgments or conclusions upon asking these questions but to complete a free association of thoughts, emotions, and mental images that come to them as well as an acknowledgment of emotional discomfort that these questions provoke, which might indicate the need to be more open and to learn from the others' perspective without necessarily agreeing with them.

F stands for showing *flexible control* in one's response to the situation. This entailed activating what is known as the social safety system, an autonomic nervous response that is associated with positive social cues like wanting to chat with others. Ernie was instructed in practicing nonverbal behaviors that signal openness to the other person, such as raising his eyebrows (otherwise known as an eyebrow wag) while engaging in a closed mouth smile and leaning back. Flexible control also involves remembering that he can delay his response and that not every situation requires immediate fixes. This was framed not as abandoning the issue but rather taking a break to reflect with a commitment to re-examine it later. Using these skills, Ernie was able to understand his parents' perspectives without giving in to them.

Another skill from RO DBT is the *Awareness Continuum* (Lynch, 2018b), which Ernie was guided in using to communicate emotions effectively to his parents, in a way that communicated his intention to take responsibility for his emotional reactions. For example, he could say, "I am aware of imagining the thought that you might be disappointed in me if I disagreed with you."

If clients are unable to get validation from others, they could use the grief protocol from RO DBT (Lynch, 2018b) to help them accept the loss of others or the world not meeting our expectations, such as for fairness and reciprocity. Another valuable tool to employ when one is unable to get validation from others is radical acceptance from dialectical behavior therapy (DBT), according to which we commit to turning toward the possibility of accepting reality as it is without trying to reject it or deny it (Linehan, 1993). For Ernie, this means actively hoping for a better relationship with his parents while not fighting the reality of what this relationship looks like at this moment.

Ernie's parents' perception of his lack of gratitude or awareness for their sacrifices is similar to intergenerational tension that can come up within the LGBTQIA+ community when queer elders, who lived through seeing friends and romantic partners die of Acquired Immunodeficiency Syndrome (AIDS) and were legally prohibited from marriage, interact with younger individuals who are able to get married and live relatively healthier. *Body Electric* (Demos, 2025), a rich and moving documentary that ties body image problems within the LGBTQIA+ community to ongoing social justice issues faced by this community, portrays how older queer individuals express happiness for younger people having more liberties, such as the legalization of gay marriage, as well as hope that younger people will respect what elders went through.

LGBTQIA+ youth have an impetus to keep fighting for rights, legal protections and inclusion, including for LGBTQIA+ elders, if for no other reason than they too will be old one day. As will be discussed further in Chapter 3, elderly people in the LGBTQIA+ community are more likely to struggle with social isolation (Murthy, 2023) and therefore are particularly vulnerable to situations that lead to having fewer resources, such as in the wake of a natural disaster or a power outage (e.g. Goldsmith et al., 2022). Elders, in turn, can serve as mentors, who deserve respect and acknowledgement for how far they have come in gaining civil rights over the past half century. They also can help queer youth feel less pressured to look perfect in order to feel that they are good enough (e.g. Demos, 2023).

LGBTQIA+ individuals have a special stake in the climate crisis and have been pivotal in climate activism. This will be explored further in Chapter 3.

WHAT ABOUT MID-LIFERS?

Middle age offers fodder for climate action in two ways: Parenthood and constructively navigating the opportunities and challenges of midlife.

Parenthood and Climate Change

Anticipation of the catastrophes that lie ahead with unmitigated global warming, and knowledge of the carbon footprint brought by each new human life, have led some young adults to opt for not having children of their own (Williams, 2021).

I can understand and validate the choice some make to not have children. At the same time, I stand by my decision to have children, not only on an emotional level but on a philosophical one, based on the sheer conviction that having children motivated me to care more about climate change.

In *Generation Dread: Finding Purpose in an Age of Climate Crisis* (2022), Britt Wray speaks to the conflict of whether or not to have children at a time of apocalyptic forecasts. She comments on alternate perspectives beyond the binary of childless spinster versus having children in a world on dialysis, such as helping to care for others' children. These alternatives resemble families sought by individuals in the LGBTQIA+ community or chosen families. She also shares how her desire to have children conflicted with stories she had heard of teens believing their parents lacked compassion in choosing to bring them into a doomed world. Wray depicts the depth of internal conflict she and her partner faced between this perspective and the anticipation of the great meaning and purpose that having children brings to many lives.

Figure 2.1 Skylar and Colby prepare a poster for me to carry in the 2023 March Against Fossil Fuels in New York City.

Photo by author.

Similar to what I experienced, ultimately bearing a child fueled her advocacy. Wray writes,

> Since I got pregnant, the tenor of my advocacy for climate justice has become even fiercer. The responsibility I feel for bringing another person into this situation turns out, not surprisingly, to spark a safeguarding energy that has more verve than before. (p. 236)

Having already had my first of two children when my own climate woes started to sink in, predictions of a resource-depleted, polluted earth instilled not only motivation to do something to address this crisis, but also pressure to protect future livelihood. Every day presents trade-offs, such as whether to allow them to take extra-long hot showers now means less clean water in their future. Some might say these trade-offs are imagined given we haven't been as affected by climate change as others at present, but more accurately, they exist in that space of unmanifested and yet likely future.

The trade-offs also roll the dice with my peace of mind. Overwhelmed with motherhood that coincided with full-time work and striving to carve out a place for myself in my career as a clinical psychologist, I chose to pass on cloth diapers for my babies, but every wet wipe used was a reminder of towering landfills.

As my children have grown, I have experienced the joy of sharing my passion for addressing the climate crisis with them. While the kids chose not to participate in the September 17, 2023 March Against Fossil Fuels in NYC, they created a poster for me to carry in the protest (see Figure 2.1). I have found allowing them to express their own environmental goals to be more successful than trying to impose tasks on them. As Linda Goldman shared (personal communication, May 23, 2025), going on walks outside and collecting leaves and rocks can strengthen their connection to nature which ultimately will sow the seeds for nurturing pro-environmental values that are unique to them. Knowledge of climate change has impassioned my children to talk about how much they care about the environment. It hasn't scared them like climate denialists have accused me of trying to do (see Climate Deniers section of Chapter 1).

Goldman (personal communication, May 23, 2025) also emphasizes the importance of understanding a child's stage of cognitive development when talking to them about the climate crisis. For example, a young child in the pre-operational stage of development might assume that effects of climate change, such as flooding, are their fault. This is a facet of early childhood cognition known as egocentrism. This is not egocentrism in a narcissistic sense, but rather in assuming that a natural phenomenon occurs because of or to serve them (i.e. the sun sets because it is time for me to sleep) (Rathus, 2022).

In contrast, an elementary school aged child can understand their capacity to do things to help solve the problem, such as participating in a beach clean-up. By adolescence, the individual might choose to join with peers to advocate for environmentalism. Even then, and into adulthood, they need not seek activism alone to have an impact. Some will pursue individual, meaningful acts like reducing food waste or producing art that portrays environmental values. Keep in mind that knowledge of cognitive development in childhood is not meant to represent strict guidelines, as individual children develop at different rates and in various contexts. Some may spearhead activism at earlier ages. Others, even into adulthood, may need more time to process the climate crisis in a healthy way that is not so overwhelming as to lead to shutting down or affecting their quality of life.

As emphasized by Wray (2022), it is important for parents to learn to cope with their own eco-distress, as children are apt to pick up on and be affected by their parent's emotions, including feeling distressed by their parents' distress. Ultimately, a developmentally sensitive, child-led approach, with a foundation laid via family time in nature, as much as is possible depending on how much green space is available to the family, can set the stage for a collaborative journey in sustainability.

MID-CAREER SHIFTS AND CLIMATE CHANGE

Research (Infurna et al., 2020) has debunked the notion that the midlife crisis is an inevitable part of the middle-aged years. Still, for some it can be a time fraught with demands, as well as concern for what kind of legacy one will leave behind.

Parenthood is a form of legacy and the human search for meaning might partly explain why those with the means to do so pour ample resources (i.e. time and wealth) into their progeny with sanctimonious zest. For other mid-lifers, there is instead, or additionally, the weight of making one's mark in their chosen career. Many pursue both, making this period of life particularly chaotic.

Kieran Setiya, a tenured philosophy professor at MIT, and author of *Midlife: A Philosophical Guide* (2017), turned to his own discipline for guidance when he found himself languishing in the "what's next?" mindset despite having accomplished all he set out to do in his younger years: a tenured faculty position at a prestigious university, a spouse and children, and multiple publications.

Setiya (2017) delves into the history of the midlife crisis, highlighting creative shifts in ancient masters, such as Michelangelo that emerged from a lack of fulfillment in their work. He also discusses the three factors that create angst for mid-lifers: Narrowing of options (starting medical school at 40, for instance, does not seem as feasible as it did 20 years ago), regrets about the past, such as the paths not chosen, and the drudgery of completing the same work again and again. Speaking further to the latter point, Setiya discusses the ennui that emerges from the sense of futility in the present, mainly in expending much of your time and energy putting out fires rather than innovating or creating.

While climate change triggers many challenging emotions, the fight has also brought meaning at different levels. It has motivated me to be a parent who understands the futility of demanding straight A's when Generation Alpha's calling will demand wisdom and flexibility, not book smarts alone (although education will also be important).

Of course, pursuing passions as a way of mitigating mid-life angst is not a new concept. Many choose to pick up hobbies, take leadership positions in their children's school board or volunteer in the community. We could wistfully proclaim that the solution to the climate crisis demands, in part, that people choose to channel these passions into climate advocacy.

But, as I will address repeatedly throughout this book, I believe that this agenda is both futile and short sided. We need to embrace the fact that climate voices can be channeled through all different passions, whether it is Jeremy Strong's decision to play a lead role in a play about environmentalism (see Chapter 1), or the medical professional who lectures to their colleagues about the health consequences of air pollution and worsening allergy seasons.

***Figure** 2.2 Skylar in front of a work of "trash art" at Duke Riley's Brooklyn Museum Exhibit, DEATH TO THE LIVING,* Long Live Trash, *December 2022.*

Photo by author.

It could be in attending Duke Riley's trash art exhibit DEATH TO THE LIVING, *Long Live Trash* (discussed by Ryzik, 2022), which I shared with my kids a few years ago in Brooklyn (see Figure 2.2). Whereas the younger generations focus more on careers in environmental science, the middle agers can unite climate change with various life domains, making it an integral part of our collective psyche.

CONCLUSION: CLIMATE CHANGE THROUGH A DEVELOPMENTAL LENS

In this chapter, I explored climate change responses and coping through the lens of generational and age differences and considered human development throughout the lifespan. I argued that intergenerational collaboration can serve to balance wisdom with innovation, expressing urgency with measured responses that might be more effective, and mitigating perfectionism. Intergenerational collaboration can also manifest in raising earth-conscious children and can be fodder for a renewed sense of purpose in middle age through community work and/or mentorship.

By aiming to open-mindedly understand the context of each other's knowledge base, cognitive perspectives (i.e. perception of time) and mechanisms for emotion regulation, we can learn to create a variegated and harmonious force to collectively pursue our passion for a cleaner and healthier planet.

REFERENCES

APA Dictionary of Psychology (Revised April 19, 2018a). *Definition of prefigurative cultures.* https://dictionary.apa.org/prefigurative-culture

APA Dictionary of Psychology (Revised April 19, 2018b). *Definition of postfigurative cultures.* https://dictionary.apa.org/postfigurative-culture

APA Dictionary of Psychology (Revised April 19, 2018c). *Definition of cofigurative cultures.* https://dictionary.apa.org/cofigurative-culture

Carstensen, L. (2021). Socioemotional selectivity theory: The role of perceived endings in human motivation. *The Gerontologist, 61*(8), 1188–1196.

Carstensen, L., & Reynolds, M. E. (2023). Age differences in preferences through the lens of socioemotional selectivity theory. *The Journal of the Economics of Ageing, 24,* 100440. https://doi.org/10.1016/j.jeoa.2022.100440

Carstensen, L., Turan, B., Scheibe, S., Ram, N., Ersener-Hershfield, H., Samaez-Larkin, G., Brooks, K. P., & Nesselroade, J. R. (2011). Emotional experience improves with age: Evidence based on over 10 years of experience sampling. *Psychological Aging, 26*(1), 21–33. https://psycnet.apa.org/doi/10.1037/a0021285

Center for Disease Control and Prevention (CDC) (2024, June 25). Heat and older adults (aged 65+). *CDC Heat health.* https://www.cdc.gov/heat-health/risk-factors/heat-and-older-adults-aged-65.html#print

Cheng, S.-T., & Yim, Y.-K. (2008). Age differences in forgiveness: The role of future time perspective. *Psychology and Aging, 23*(3), 676–680.

Cigna Health and Life Insurance Company or Their Affiliates (2022). *Cigna 360 global well-being survey: Exhausted by work-the employer opportunity.* Cigna.

Cruikshank, M. (2013). *Learning to be old: Gender, culture, and aging.* Rowman & Littlefield Publishers.

Curran, T., & Hill, A. P. (2019). Perfectionism is increasing over time: A meta-analysis of birth cohort differences from 1989 to 2016. *Psychological Bulletin, 145*(4), 410–429. https://doi.org/10.1037/bul0000138

Dekornfeld, O. (2021). "Teenage brain" episode of *The Mind Explained* (Season 2, Episode 2) [Netflix documentary series].

Demos, N. (Director) (2023). *Body electric* [Documentary film]. Nick Demos Media.

Dimock, M. (2019, January 17). *Defining generations: Where Millennials end and Generation Z begins.* Pew Research Center. https://www.pewresearch.org/short-reads/2019/01/17/where-millennials-end-and-generation-z-begins/

Dimock, M. (2023, May 22). *Five things to keep in mind when you hear about Gen Z, Millennials, Boomers, and other generations.* Pew Research Center. https://www.pewresearch.org/short-reads/2023/05/22/5-things-to-keep-in-mind-when-you-hear-about-gen-z-millennials-boomers-and-other-generations/

Dreyfus, J. L. (Host). (2023, May 16) Julia gets wise with Gina McCarthy (Season 1, Episode 7) [Audio podcast episode]. In *Wiser than me.* Lemonada.

Francioli, S., & North, M. (2021). Youngism: The content, causes, and consequences of prejudice toward younger adults. *Journal of Experimental Psychology: General, 150*(12), 2591–2612. https://doi.org/10.1037/xge0001064

Franco, C., & Ghisetti, C. (2022). What shapes the "value-action" gap? The role of time perception reconsidered. *Economia Politica, 39*, 1023–1053. https://doi.org/10.1007/s40888-022-00282-8

Goldman, L. (2022). *Climate change and youth: Turning grief and anxiety into activism* (1st ed.). Routledge.

Goldsmith, L., Raditz, V., & Méndez, M. (2022). Queer and present danger: Understanding the disparate impacts of disasters on LGBTQ+ communities. *Disasters, 46*(4), 946–973. https://doi.org/10.1111/disa.12509. Epub 2022 Aug 9. PMID: 34498778.

Goldstein-Rose, S. (2020). *The 100% solution: A plan for solving climate change.* Melville House.

Goodall, J., & Abrams, D. (2021). In J. Goodall and D. Abrams (Narrs), *The book of hope: A survival guide for trying times* [Audiobook]. Celadon Books.

Harvey, A. S., & Mukhopadhyay, A. R. (2007). When twenty-four hours is not enough: Time poverty of working parents. *Social Indicators Research,* 82, 57–77. http://www.jstor.org/stable/20734446

Hickman, C., Marks, E., Pinkala, P., Clayton, S., Lewandowski, E., Mayall, E. E., Wray, B., Mellor, C., & Van Susteren, L. (2021). Climate anxiety in children and young people and their beliefs about government responses to climate change: A global survey. *The Lancet Planetary Health, 5*(12), e863–e873.

Hirsh, S. (2023, February 9). Merlin van Lawick on how his grandma, Jane Goodall, inspires him: "She represents hope" (Exclusive). *Greenmatters.* https://www.greenmatters.com/environmental-leaders/merlin-van-lawick

Höfrová, A., & Small, M. A. (2024). A systematic literature review of education for generation alpha. *Discover Education, 3*(125). https://doi.org/10.1007/s44217-024-00218-3

Imolore, D. (2023, August 25). Environmental efforts for global climate impact. *Fund the Planet.* https://blog.fundtheplanet.net/fund-the-planet/micro-sustainability-individual-environmental-efforts-for-global-climate-impact/

Infurna, F. J., Lachman, M. E., & Gerstorf, D. (2020). Midlife in the 2020s: opportunities and challenges. *American Psychologist, 75*(4), 470–485. http://dx.doi.org/10.1037/amp0000591

IPCC. (2023). Summary for policymakers. In H. Lee & J. Romero (Eds.), Climate change 2023: Synthesis report. contribution of working groups I, II and III to the sixth assessment report of the intergovernmental panel on climate change [core writing team.]. IPCC, Geneva, Switzerland, pp. 1–34, https://doi.org/10.59327/IPCC/AR6-9789291691647.001

Jennett, T. (Director). (2022). *Gen Z mental health: Climate stories! [film].* Astronaut Productions.

Keluskar, J. (2024, June 29). All-for-nothing angst. *Alone Together.* https://www.psychologytoday.com/us/blog/alone-together/202406/all-for-nothing-angst

Kurzban, R., Duckworth, A., Kable, J. W., & Myers, J. (2013). An opportunity cost model of subjective effort and task performance. *Behavioral and Brain Sciences, 36*(6), 661–679. https://doi.org/10.1017/S0140525X12003196

Lawson, D. F., Stevenson, K. T., Peterson, N., Carrier, S. J., Strand, R. L., & Seekamp, E. (2019). Children can foster climate change concern among their parents. *Nature Climate Change, 9*, 458–462. https://doi.org/10.1038/s41558-019-0463-3

Lewandowski, R. E., Clayton, S. D., Olbrich, L., Sakshaug, J. W., Wray, B., Schwartz, S. E. O., Augustinavicius, J., Howe, P. D., Parnes, M., Wright, S., Carpenter, C.,

Wiśniowski, A., Ruiz, D. P., & Van Susteren, L. (2024). Climate emotions, thoughts, and plans among US adolescents and young adults: A cross-sectional descriptive survey and analysis by political party identification and self-reported exposure to severe weather events. *Lancet Planet Health, 8*, e879–93.
Linehan, M. (1993). *Cognitive behavioral treatment of borderline personality disorder.* Guilford.
Lynch, T. (2018a). *Radically open dialectical behavior therapy: Theory and practice for treating disorders of overcontrol.* Context Press.
Lynch, T. (2018b). *The skills training manual for radically open dialectical behavior therapy: A clinician's guide for treating disorders of overcontrol.* Context Press.
McCauley, E. (2023, August 11). A hidden currency of incalculable worth. *The New York Times.* https://www.nytimes.com/2023/08/11/opinion/time-poverty-money.html
Munchies. (2015, March 30). *How to make chicken cutlets with Frank Pinello & his Nonna* [Video]. YouTube. https://www.youtube.com/watch?v=wttzz0RtMsU&t=2s&ab_channel=Munchies
Murthy, V. (2023). *Our epidemic of loneliness and isolation: The U.S surgeon general's advisory on the healing effects of social connection and community.* US Public Health Service.
Pipher, M. (2023, April 16). Grandmothers of the world, unite. *The New York Times.* https://www.nytimes.com/2023/04/16/opinion/climate-change-grandmothers.html
Rathus, S. (2022). *Childhood and adolescence: Voyages in development* (7th ed.). Cengage.
Robinson, A., Divers, R., Moscardini, E., & Calamia, M. (2021). Perfectionism, conscientiousness, and neuroticism: Does age matter? *Personality and Individual Differences, 172*(7). https://doi.org/10.1016/j.paid.2020.110563
Romero, A. (2020, September 18). In memory of Justice Ruth Bader Ginsburg (1933–2020). *ACLU News and Commentary.* https://www.aclu.org/news/civil-liberties/in-memory-of-justice-ruth-bader-ginsburg-1933-2020
Ross, A. (2020, April 21). 5 great things we should never forget about Ruth Bader Ginsburg (1933–2020). *Good News Network.* https://www.goodnewsnetwork.org/5-great-reasons-to-remember-ruth-bader-ginsburg/
Ryzik, M. (2022, June 16). Duke Riley: Grand master trash. *The New York Times.* https://www.nytimes.com/2022/06/16/arts/design/duke-riley-artist-brooklyn-museum-trash.html
Setiya, K. (2017). *Midlife: A philosophical guide.* Princeton University Press.
Vendantem, S. (Host). (2023, June 27). The best years of your life (Season, Episode) [Audio podcast episode]. In *Hidden brain.* Hidden Brain Media.
Williams, A. (2021, November 20). To breed or not to breed? *The New York Times.* https://www.nytimes.com/2021/11/20/style/breed-children-climate-change.html
Wray, B. (2022). *Generation dread: Finding purpose in an age of climate crisis.* Knopf Canada.

Chapter 3

Understanding Your Climate Identity

Identity is a framework for understanding ourselves, as well as a lens through which we interpret others' behaviors. In the face of existential threats such as the climate crisis, we can join with others who we perceive to share similar facets of identity. On the other hand, perceiving discrepant identities in those we interact with can lead to "othering," a process that leads to assumptions, generally negative, about those whom we perceive as different (e.g. Guzmán, 2022).

Identity formation is a lifelong process that prompts questions: Who am I, and what is my place in the world? What mark can I leave in the blip of time I'm granted on earth? Various aspects of my identity guided me in finding a productive role in environmental advocacy. I joined climate advocacy groups for parents and mental health providers. While I found commonality with the members of these groups, other aspects of my identity made it difficult to join with others. For example, my meekness didn't scream "driver of social transformation."

I wanted to do something meaningful to help, but being able to organize as an activist, or to convince others to join my cause, did not seem to be part of my skill set, as it isn't for many who struggle to be assertive. Time and again, my fight against climate change turned more into a fight against the least appealing aspects of myself.

Identity can be conceptualized in individual and collective dimensions (Nevid, 2022), both of which are relevant to joining together in the environmental cause.

DOI: 10.4324/9781003489665-4

COLLECTIVE IDENTITY

Collective identity refers to how we see ourselves as part of and in relation to a larger group (Nevid, 2022). According to Joanna Macy and Chris Johnstone, co-authors of *Active Hope: How to Face the Mess We're in with Unexpected Resilience and Creative Power* (2022), a shift toward collective identity allows us to transcend the hierarchies of wealth and status that humans have created.

While collective identity generally applies to large, diffuse groups, such as your hometown, nation, school, workplace, or family, it can also describe goal-oriented groups that come together for a cause. Some climate advocacy groups, such as Moms Clean Air Force, the Sunrise Movement (for youth), GASP for change (a grandmother's group), and OUT for Sustainability (an environmental group geared toward the LGBTQIA+ community) align with an additional aspect of identity. In the above examples, it is an aspect of identity that interacts with climate change to create a sense of personal threat, such as a parent group's concern for how environmental devastation affects their children.

The benefit of linking environmental collectivism with other aspects of personal identity is that it motivates people to share with environmentalism the time and resources they devote to other highly valued areas. For example, some parents might give up hobbies because they see these as self-serving when time and resources would be better spent on their children. However, if they see climate activism as part of their parental duties, there are fewer qualms about putting efforts toward it. Linking environmental work with career goals, as evidenced by me writing this book, is another example of this idea.

Unfortunately, we live in a world in which pursuing goals often requires actions that run counter to environmentalism, seemingly leaving you forced into a binary of either pursuing the goal or pursuing environmentalism. For example, career success has been linked with making as much money as possible, limiting time for creative pursuits that could have a sustainable theme. As mentioned previously, perfectionism has increased in college students over the past four decades (Curran & Hill, 2019), which means more time is spent on achieving what is thought to be important for a career and less time is spent on learning for the sake of learning and self-exploration, including exploring one's values and engaging in voluntary advocacy work. Being a good parent means affording to pack your child's schedule with activities and purchase enough toys, gadgets, and vacations to keep them content. Striving to be a good-enough everything, let alone, a good-enough anything is pushing us to put sustainability on the backburner.

Finding an environmental group that aligns with an additional aspect of your identity affirms your choice to participate in environmental advocacy. If it is OK for a group of moms to write to petition their children's school district for electric school buses, then it feels more appropriate for me to join in their efforts.

However, the downsides of these collectives are the potential for siloing and exclusion of people who don't fit the identity tied to the groups, such as people who aren't parents not feeling like they fit in with parent-based environmental groups.

My goals in this chapter are: (1) describe how psychological and social identity intersect with the climate crisis in the hopes of underscoring the importance of guiding individuals to find their unique role in advocacy, one that will be strengthened through a collective identity; and (2) to discuss the idea of coalition building among groups representing various identities.

The first step is to understand domains in which people find identity, the second to appreciate these identities, and the third is to help them unite in solidarity. I will guide the reader through all three steps with practical suggestions. I will do this through the lens of various aspects of identity: Personality, gender identity and sexual orientation (LGBTQIA+, or Lesbian, Gay, Bisexual Transgender, Queer/Questioning, Intersex, and Asexual, with the "+" representing additional identities), racial/ethnic, socio-economic, political and ideological, and neurodivergence.

PERSONALITY

Personality can be defined as the relatively stable set of psychological characteristics and behavior patterns that make individuals unique and that account for the consistency of their behavior over time (Nevid, 2022). Personality is one of the catchier areas of psychological fields, ranging from Freud's theory of psychosexual development, positing that our typical behavioral repertoire is shaped by unconscious forces and sexual impulses, to humanistic theories which emphasize desires to attain our potential, also known as self-actualizing.

Trait theory is a particularly influential perspective of personality, with the Five-Factor Model or the "Big Five" being the most heavily researched (Nevid, 2022). These Five Factors include Openness to experience, Conscientiousness, Extraversion, Agreeableness, and Neuroticism, all of which exist on a continuum. Openness includes being open to different perspectives and novel experiences, such as trying a new cuisine. Conscientiousness speaks to norm-following or doing what is deemed good, whether it is about health, achievement, or morality, and it implies a tendency to persist in goals and delay gratification. Extraversion refers to the extent to which one is outgoing and thrives emotionally and socially from interacting with others. Agreeableness, like openness , to experience not only requires a degree of psychological flexibility, a willingness to go with the flow, but also entails friendliness, compromise, and deference.

The main limitation of trait theory is important to keep in mind, and that is its circular nature (Nevid, 2022). For example, let's say we describe Jenny as friendly. What makes Jenny friendly, you might ask? Trait theory would describe certain characteristics associated with this trait, such as, "She tends to smile at people and make polite conversation. She offers compliments, etc." If you were to then ask why Jenny acts in these ways, trait theory would say "Because she's friendly."

Nevertheless, the fascinating and relevant aspect of trait theory research is in its predictive quality. For instance, research indicates that conscientious people tend to engage in healthier behaviors and have biomarkers of better health, and neurotic people tend to experience poorer coping with stress (e.g. Clague & Wong, 2023) and overall worse health outcomes (e.g. Milad & Bogg, 2020; Yoneda et al., 2023). While there are numerous factors, both genetic and environmental that impact health outcomes, these associations are alluring in the sense of sparking the question, how do our ways of interfacing with the world impact our destinies?

Research has shown that 50% of the variability we see in people's personality traits is accounted for by genetic differences, while the other 50% is accounted for by shared environment, or factors within the family's social dynamics as well as unshared environment, which includes both social experiences that are unique to the individual and random occurrences in development such as exposures in utero that influence the expression of genes (Linden, 2020). In addition, different contexts can bring out varying aspects of personality. In *Unique: The New Science of Human Individuality*, David Linden (2020) points out that while the stereotype of the first-born child being socially dominant might hold true in the child's home environment, there is no evidence of this association in environments outside of the home, as the child may no longer enjoy the same status they had within their family.

The individual's social world, therefore, brings out different aspects of personality and selectively reinforces them. For example, individuals with generalized anxiety disorder (GAD) tend to overestimate their affiliation with others, falsely believing, for instance that others should always reciprocate things they do to strengthen a social bond, when ironically, rigidly adhering to this belief fuels hostility when the other person doesn't fulfill their expectations (Shin & Newman, 2019). This leads to them being more likely to get exploited by others, fueling a vicious cycle and one that ultimately leaves individuals with GAD feeling less connected with others and feeling less empowered by their relationships.

This is important to note because research has found that anxious people with higher environmental values are more likely to engage in climate action (Pickering & Dale, 2023). Another study, by the same authors, found that the strongest psychological predictor of inaction was a sense of powerlessness (Pickering & Dale, 2024). While predictors of action also included knowledge about climate change, its severity, and the perceived urgency to act, powerlessness was a stronger predictor than these factors. Therefore, we can't assume that knowledge and urgency to act are sufficient in motivating climate actions. One must negotiate the relative strengths and limitations that their personalities present in interacting with their chosen role to make it successful and fulfilling.

One (De Jong et al., 2019) analysis offers a clarifying perspective about how personality predicts one's behavior in the workplace. The researchers found that three of the Big Five Traits, Extraversion, Conscientiousness, and Openness contribute to career role preferences, including what the authors refer to as *Maker, Expert, Presenter, Guide, Director* and *Inspirer* role preferences. These role preferences mediate or account for the relationship between Big Five traits and preferred roles within a chosen career. Although this study was cross-sectional, meaning it offered only a snapshot of the participants as opposed to being a longitudinal study that followed the participants over time, the level of depth offered in their analysis renders it worthy of exploring here.

For instance, *Makers* (De Jong et al., 2019) strive for personal mastery and success, and value autonomy and independence. They thrive in work environments with clear expectations and guidelines. *Experts* also value independent agency and show enthusiasm for exploratory problem solving. In contrast, *Presenters* thrive on interacting with and influencing others. The *Guide* role similarly emphasizes interactions with others but in a way that focuses on understanding their perspectives as opposed to trying to influence them. The *Director* role emphasizes attaining long-term goals from a position of dominance. The *Inspirer* role, like the Director's, focuses on optimizing strategy, but is more concerned with initiating change than with dominance.

Tying this back to the Big Five traits, the study (De Jong et al., 2019) revealed Conscientiousness to be related to the *Maker* role, Openness to Experience to positively predict preference for the *Expert* role, *Extraversion* to predict preference for the *Presenter* and *Director* roles, and Agreeableness/Friendliness to be associated with the *Guide role.* Both Extraversion and Openness to Experience were related to preference and engagement in the *Inspirer* role.

We can extend the idea of role preferences to help climate advocates determine the most fulfilling role they can play in climate action. For example, one study (Zawadzka et al., 2024) found that both an intrinsic goal, community feeling, and a materialist goal, popularity were positive predictors of activist identity and commitment. Given that popularity, by definition, is exclusive, it makes sense to focus on helping people find their unique role within a community in order to help them build a stronger identity in relation to that community. While the roles described by De Jong et al. (2019) emphasize that success can be achieved in multiple ways, not by popularity alone, there is no guidance in identifying aspects of personality that might interfere with achieving those roles.

Adam Grant, an influential organizational psychologist devoted to helping people overcome such obstacles, co-created *Principles You* with Ray Dalio, Brian Little, and John Golden (Dalio et al., 2021). *Principles You* is a free online personality inventory that can be completed individually or as a team (www.principlesyou.com). The assessment has been researched and aligns with the Big Five traits. Based on the results of the inventory, you are placed into 1 of 28 archetypes, along with 2 runners up that are close matches. The 28 archetypes are grouped into 10 categories: Leaders; advocates, enthusiasts, givers, architects, producers, creators, seekers, fighters, and individualists. Each of these 10 categories includes 3 archetypes except for Individualist, which stands on its own.

Upon completing the inventory (www.principlesyou.com), my results placed me as being most like an Individualist, the one on its own island, described as someone whose strengths include creativity, thinking differently, and being sensitive to others' emotion. Areas for growth include coping with moodiness and self-doubt, which could get in the way of forging important relationships in the workplace, and learning to not take constructive feedback personally. Individualists can be emotionally reactive when others don't see things the way they do.

While the results of inventories like these are far from deterministic, they have given me reason to self-reflect on frustrations I have faced with collective action. It's validating to know that being described as a personality type that exists on its own island might explain why I have had trouble with group work.

The idea of using a trait assessment to guide people in finding their role in environmentalism is explored extensively by Heather White, author of *One Green Thing* (2022). White developed a Service Superpower Assessment based on 20 years of experience in the field of environmental advocacy. Her purpose is to guide people in finding pro-environmental behaviors that are more likely to stick because they are in line with the individual's values and identity. White provides descriptions for each Service Superpower as well as examples of actions that might align with them. She even includes a 21-Day Kickstarter Plan for each type.

For example, an *Adventurer* (White, 2022) values outdoor, hands-on learning and the physicality of the outdoors, which might align with organizing a family hike or speaking to the community about the benefits of time outside. *Beacons* have a strong sense of justice and might do well by writing letters to congress or recruiting friends for a cleanup. An *Influencer* shares information and events with friends and can encourage friends to use more sustainable brands. A *Philanthropist* donates time and/or money to the cause, such as by volunteering at a local nature center. *Sages* contemplate the spiritual connection to nature and might organize a community lecture on faith and nature. A *Spark* is likely to agree to attend climate awareness events. A *Wonk* loves science, policy, data graphs, and charts and would do well sharing favorite documentaries with friends and writing an op-ed in the local paper about drinking-water pollution.

The pros of White's (2022) work are that her guide is user-friendly, has the potential for popular appeal in assigning people an identity with associated practical roles, and offers a range of advocacy and sustainability behaviors that range in time and effort intensity. In other words, she reinforces the notion that climate advocates do not need to fit a single mold. However, there are limitations to her tool. While White has knowledge and experience in environmental science and advocacy, her inventory has not been subject to the same scientific rigor that social scientists would use.

However, White's inventory is not necessarily problematic as long as we know how to interpret it in the correct context: As a theoretical guide as opposed to a validated personality assessment. This framework allows for more flexibility than a personality type would suggest. For example, I completed her inventory two times within a one-year time period with different results each time, which might speak to how I have evolved in my climate work as opposed to having only one type that defines me. Personality traits, in contrast, tend to remain stable (that is they don't change much) throughout adulthood (Nevid, 2022).

Introversion–Extraversion Continuum

While being armed with knowledge about how your personality can be helpful in the fight against climate change can empower people, it can still be difficult to overcome assumptions we hold about ourselves.

One assumption is that introverts are not as powerful at inspiring social change. The results of research extolling extraversion strengthen this assumption, such as a finding from one of the studies discussed above that extraversion was positively related to climate change proactivity (Cipriani et al., 2024). Does this mean that introverts, those who might tire easily from the networking inherent in influencing change, or who might be too shy to march and shout at the doors of government offices or chain themselves to the doors of banks, are to be left out of the fight for climate justice?

According to introverted climate activist Sarah Corbett, the answer is a most definite "No." Corbett did a Technology, Entertainment, and Design (TED) talk (2016) that has had over 1.3 million views on TED's online platform, in which she discussed not only how introverts and highly sensitive people can be and should be included in climate activism, but how people with these traits offer special advantages to the cause as well. She emphasizes that we are missing the power of a large segment of the population by not including them in activism.

As Corbett (2016) explains, activists tend to react quickly, while introverts can shine in their potential to engage in a quiet, slow, and intimate form of activism, which she recommends expressing through creative works, such as crafting. Corbett founded the *Craftivist Collective*, whose mission is to work with introverts and Highly Sensitive People, people struggling with anxiety, and those who want to fight for a cause but don't know what to do. Her idea came out of her desire to find an alternative to conventional activism, from which, by the age of about 25, she felt burnt out from the amount of confrontation it demanded (Corbett & Housley, 2011; Craftivist Collective, n.d.). As her story goes, Corbett decided to make a handkerchief for a politician who had been ignoring her petitions, on which she embroidered the message "Don't blow it ..." to mean that she was encouraging the politician not to miss the chance to use her position of power for positive change. Corbett emphasizes that such gentle activism sends the message to politicians that activists are "critical friends" as opposed to adversaries.

In advocating for introverted and emotionally sensitive activists, Corbett is essentially calling for social justice in social justice. Along the lines of catering to different personalities, it is important to consider how individuals who think, behave, and process information and emotions differently might respond to the climate crisis in different ways. I next discuss individuals who identify as neurodivergent, with a focus on autistic individuals.

NEURODIVERGENCE AND THE CLIMATE CRISIS

> There *are* some issues that are black and white. There are indeed planetary and societal boundaries that must not be crossed
>
> (Thunberg, 2022, p. 1)

Greta Thunberg is the 23-year-old renowned leader of the global movement *Fridays for a Future*, which she started when she was 15-years-old by sitting out of school to protest the government's inaction to address climate change (BBC, 2024). Prior to launching into climate action, she was introverted, struggled with mental illness, and was diagnosed with Asperger's

Disorder (now referred to as autism spectrum disorder, ASD) around the age of 11- or 12-years-old. Her engagement in the climate crisis helped her overcome her emotional struggles. Thunberg describes autism as her superpower (BBC, 2024), as it has led her to see through the smoke of disingenuous politicians and corporations, and, in thinking differently, to take on novel approaches such as the school strikes.

Part of thinking differently might be the tendency to engage in literal, black, and white thinking. Such an all-or-nothing thinking style is disadvantageous in many contexts, but climate change is one of the few where it might be adaptive. As Solomon Goldstein-Rose explains in *The 100% Solution: A Plan for Solving Climate Change,*, partially addressing the climate crisis will be insufficient. Just like youth, as explained in Chapter 2, serve the role of pushing older generations into action, black and white thinkers can push the neurotypical to understand that, no, they are *not* over-reacting about the enormity of this crisis.

Early in my career, I worked with several children who were similar to Greta, very intelligent and sensitive neurodivergent youth. Equally important as using evidence-based interventions was showing up for them with genuine transparency, a desire to celebrate their strengths, and compassion for the challenges that come with navigating a world that does not prioritize authenticity as much as they do.

While some of the youth I worked with shared a passion for social justice, I also worked with autistic youth who denied climate change. Consistent with my varied clinical experience, research has failed to show a relationship between Autism and climate change concern. In contrast, Taylor et al. (2021) measured autistic traits and environmental beliefs and behaviors in undergraduate students and found that self-reported autistic traits were associated with engagement in *fewer* pro-environmental behaviors and predicted neither pro-environmental attitudes nor climate change belief. In other words, having more autistic traits did not make the participants more likely to hold pro-environmental values nor any more likely believe in climate change. Rather, having these traits made them less likely to engage in pro-environmental behaviors.

Based on these results (Taylor et al., 2021), the researchers concluded that people should "move away from autism-based narratives" of climate activism and focus more on other aspects of neurodivergent environmentalists' identities, such as younger age.

However, autistic individuals comprise a diverse group. There is also the possibility that if we broaden the scope from ASD specifically to the ever-growing category of neurodivergence, which encompasses ASD, Attention-Deficit/Hyperactivity Disorder, learning and other developmental disorders (Gray et al., 2025), we might see a clearer trend toward greater desire to engage in climate justice. More research is needed in this area.

Neurodivergence is associated with co-morbid, or co-existing mental health issues such as mood disturbance and anxiety (e.g. McKenney et al., 2024). When we consider that anxiety traits are related to climate change concern and pro-environmental behavior (Pickering & Dale, 2023), and that sensory sensitivity, such as being overwhelmed by strong sensory input, is another trait of neurodivergence that is also associated with a connectedness to nature and animals (Setti et al., 2022), it makes sense to consider that aspects of neurodivergence can be an asset for fueling motivation to fight for a cleaner planet.

In addition, there is a plethora of literature pointing to the vulnerabilities of neurodivergent individuals to the effects of the climate crisis, so they have a stake in the game. For example, some psychiatric medications make people more vulnerable to the negative health effects of heat (West & Yockey, 2023).

Researchers such as Watfern and Carnemolla (2025) have noted that individuals with intellectual disability may lack access to resources and services in the wake of a natural disaster or health-threatening event such as a heat wave and are more susceptible to being victimized by acts of violence.

Schleck and Ben-Alon (2024) define eco-ableism as discrimination against individuals with disabilities that arises from activism, such as guilting people for using plastic straws for convenience (Pledl, 2021 as cited in Schleck and Ben-Alon, 2024). This is in line with the notion that, because our post-industrial society is set up to favor less sustainable options as being more convenient, it is important to acknowledge the privilege that dovetails sustainable behaviors. Driving electric vehicles means you can afford them, using cloth napkins means you have the time to launder them, volunteering in environmental activism means you have knowledge of how to engage in it, as well as time and energy that could otherwise be spent engaging in paid work or completing unpaid but required household duties like cooking meals for your family. Criticizing an individual with a disability for eating a frozen meal when grocery shopping and food preparation is challenging for them is an act of exclusion.

Differences in emotion regulation, such as reacting more intensely to emotional stimuli and having more trouble self-soothing, are common in autistic individuals (Mazefsky & White, 2014) and other neurodivergent individuals. People who are more sensitive emotionally including those who identify as a Highly Sensitive Person (HSPs) (Lionetti et al., 2018) might make resilience more challenging in the face of disaster.

Also, emotional stress might impede pro-environmental behavior because when we are stressed, we tend to do what is quick and easy, and all of us born in the post-industrial era are primed to deviate to less sustainable behaviors as the easier option. For example, when I'm rushed to leave a store, and I realize I left my canvas bags at home, it's quicker and easier to accept plastic bags than to place groceries individually into the car or put mental energy into thinking of another solution.

Another example of a processing difference that could either facilitate or hold you back from pro-environmental behavior is disgust sensitivity, described below.

DISGUST SENSITIVITY

As I drove my kids to school one morning, a bus passed by us right after we had opened the car windows. I braced myself as I noticed exhaust billowing from its backside. The stench hit me right away. "I hate that we don't have electric school buses, and we have to inhale this dirty air!" I complained aloud. This ignited my kids' curiosity, and a conversation about environmentalism ensued.

This is an example of how disgust sensitivity can lead to pro-environmental attitudes, but it can also have the opposite effect, such as using two forks because you are put off by using the same one for your syrup-drenched pancakes and fruit. It can fuel both sustainable and unsustainable behaviors depending on the situation. Thinking about garbage being added to a smelly landfill could motivate you to reduce waste, while being bothered by a tiny splotch on your shirt could lead to using extra resources to make it perfectly clean.

Disgust is a universal human emotion that protects us from pathogens, such as those in rotting meat by motivating us to avoid it (Young, 2021). It might also remind us of our mortality (i.e. that someday we will become rotting meat). Disgust keeps us in line as well by encouraging moral purity.

However, disgust sensitivity exists on a continuum, in that people vary widely in terms of how prone they are to feeling disgust and in what category (i.e. moral versus physiological) they experience it. As I discuss in a *Psychology Today* blog post (Keluskar, 2024), individual differences in disgust sensitivity can fuel misunderstandings. People can be put off by others who eat food that grosses them out, while the person eating it could be put off by others' sensitivity.

One suggestion is to talk openly about disgust sensitivity. A couple of summers ago, I questioned my children about why they weren't re-using their washcloths and learned that the reason was they felt grossed out by that. Talking openly about this led to problem-solving more sustainable options, such as giving each washcloth a quick manual wash with hand soap and water, which made it more acceptable for them for re-use.

There is no venue for talking openly about and respectfully about our differences that is more challenging than the thorny arena of political perspectives. This deserves a place in this book because whether we like it or not, climate change has become political and the politics surrounding it represent a formidable barrier to collective action.

POLITICAL AND IDEOLOGICAL DIFFERENCES

The results of a 2022 Pew Research Center analysis revealed that the ideological divide along Democratic (left or liberal leaning) and Republican (right or conservative leaning) party lines is the greatest it has been in the past 50 years (Desilver, 2022), and the increased divergence in ideology has been largely driven by Republicans' shift to the right.

Climate change beliefs are associated with both political affiliation and ideological orientation, but they also interact with other demographic factors. Interestingly, how they interact with these factors also varies by political party (Federico & Malka, 2018). While wealthier and more educated Democrats show greater concern about climate change, the reverse relationship is found among Republicans (wealthier and more educated Republicans show less concern about climate change).

Another line of research investigates whether certain personality traits are associated with ideological leanings. According to Federico and Malka (2023), high levels of psychological rigidity, threat sensitivity, conscientiousness, and need for closure predict right-wing ideological preferences. People who lean toward conservative ideology tend to favor strengthening communal social ties in the face of uncertainty. In contrast, openness to experience is correlated with a liberal, or left-leaning preference.

David Brooks (2024) explains that while Republicans can be rugged individualists with economics, such as favoring tax cuts for the wealthiest Americans, liberals are rugged individualists when it comes to morality, a concept Brooks refers to as the privatization of morality. This means that liberals believe that the individual decides which morals to uphold. However, this is far from deterministic, as ideological preference also depends on how messages are presented to the public. As Brooks points out, people in general, regardless of their political or ideological affiliation, want to be part of a community with a cohesive set of moral values to adhere to together. Brooks argues that this discrepancy, between the human need for communal morals and the liberal ideological preference for individualized morality, drives existential angst among left-leaning individuals.

However, this begs the question, how do we create social cohesion when our ideals about moral order are in diametric opposition? Binary thinking can be seen on both extremes of the political spectrum (Moss & O'Connor, 2020). This goes against unification and is problematic for various other reasons.

Working with anti-vax families who homeschooled their children to avoid mandatory vaccinations for matriculation at public schools taught me that you will gain more ground with emotional appeals than logic. For example, lecturing them about the solid scientific evidence that the benefits of vaccination outweigh rare risks tended to fall on deaf ears. Instead, I highlighted the emotional detriment to children, particularly those with emotional and behavioral challenges, when they lacked the structure and socialization provided by schools. I learned to be cautious about the language I used, as any scent of judgment turned them away.

As Mónica Guzmán (2022) notes in her book *I Never Thought of It That Way: How to Have Fearlessly Curious Conversations in Dangerously Divided Times*, people rarely change their minds because of pressure from someone else, but rather this change happens when they tell themselves new stories that you can craft by aligning with the aspect of them that you see in yourself. Guzmán is a staff member of a national organization called Braver Angels, which organizes debates in which participants are asked to present both sides of a controversial issue, such as when Covid-19 restrictions on businesses should be lifted.

In my case of working with anti-vax families, aligning with them as a parent allowed me to empathize with the weight of making health decisions for your child, and the distrust that can easily be engendered toward larger social entities. Laying the groundwork for this conversation by first spending time getting to know the parents and building an alliance was far more effective than talking at them from the perspective of a professional, a mistake I made earlier in my career.

Other evidence that change can emerge from relationships is shown in research suggesting family members can be agents of changing attitudes, such as one study that found that tween and teen girls were particularly effective at convincing their conservative fathers to shift their attitudes to being more concerned about climate change (Lawson et al., 2019). In turn, family values (Flanagan & Tucker, 1999) shape children's goals for materialism versus public interest. One can make a case for the solution to the climate crisis involving both decreasing materialism and nurturing social bonds.

I conclude this section on political views with a recent news story that serves as yet another example of the power that relationships can have in changing ideology. Largely pro-Trump inhabitants of a small town in Missouri were devastated when a beloved immigrant named Ming Li Hui, or "Carol," the name she went by, was detained by federal immigration officials as part of their crackdown (Healy, 2025a). Carol had worked hard for 20 years cleaning homes and serving waffles at a local diner. This shows that a relationship can cause a belief shift even if it is not a close relationship, such as a spouse or child. When people missed Carol's warmth, they had a hard time reconciling their feelings toward her with their support for Trump. About one week later, Carol was released from detention (Healy, 2025b), thanks to her community coming together to bring attention to her story.

GENDER IDENTITY AND SEXUALITY (LGBTQIA+)

Sexual and gender minorities, including Lesbian, Gay, Bisexual, Transgender, Queer/Questioning, Intersex, Asexual and all subsects individuals, whose representation in the US population has climbed to 7% (Brown, 2023), are more likely to struggle with chronic disease and mental health difficulties due to discrimination and the shame that society confers on individuals who do not identify as cishet (heterosexual individuals whose gender identity matches their assigned sex at birth).

It is not surprising that queer individuals are also more likely to endorse eco-anxiety than cishet individuals. In their paper "Queering Environmental Justice: Unequal Environmental Health Burden on the LGBTQ+ Community," Goldsmith et al. (2022) provide a compelling argument for integrating environmental justice with LGBTQIA+ advocacy.

For one, it is important to consider the relevance of intersectionality, or the effect of having multiple marginalized identities, to the LGBTQIA+ community. For example, about 40% of LGBT adults are people of color (The Williams Institute, 2021). Of these individuals, about 12% identify as Black. According to the National Survey on LGBTQ Youth Mental Health (The Trevor Project, 2021), 75% of LGBTQ youth reported being discriminated against at least once in their lifetime for their sexual orientation or gender identity, and half of all LGBTQ youth of color reported being discriminated against for their race/ethnicity in the past year. This included 67% of Black LGBTQ youth and 60% of Asian/Pacific Islander LGBTQ youth. Individuals with Autism, who have been subject to ableism, are also more likely to identify as gender non-conforming (Stokes & Stokes, 2018).

Communities that have historically been discriminated against are more likely to be the victims of violence in the wake of a natural disaster, and to be excluded from humanitarian aid (Carter, 2021). This has included gender-based violence towards LGBTQ+ individuals and women (Roure, 2019). In addition, sexual and gender minorities are more likely to be unhoused and to report financial concerns (Sears, 2024) which also render them more vulnerable to exposures of disasters and barriers to recovery. The affiliation of aid organizations with faith-based institutions increases the risk for anti-LGBTQ+ discrimination as well as rendering sexual and gender minorities more hesitant to seek assistance from groups that have historically discriminated against them.

Goldsmith et al. (2022) also describe many potential environmental exposures to which sexual and gender minorities are particularly prone, including respiratory distress exacerbated by the interaction of air pollution and chest binding, and immunological vulnerabilities due to HIV infection and other chronic health problems. These vulnerabilities underscore the need for destigmatizing HIV, access to gender affirming care, and access to medicines, which could be difficult to access in polluted environments and in the wake of disasters. Mental health disorders and stress are also more common in sexual and gender minorities due to systemic factors and can be further exacerbated by extreme heat and other environmental hazards.

In another paper by Goldsmith et al. (2021) entitled "Queer and Present Danger: Understanding the Disparate Impacts of Disasters on LGBTQ+ Communities," the authors make the case that LGBTQIA+ individuals have been excluded from disaster planning and unrecognized within US disaster and response policies and disaster programs. They cite the increased risk of poverty and social isolation, particularly among elderly queer individuals (Goldsen, 2018), as well as higher rates of incarceration among LGBTQ+ individuals (e.g. Cain & Ellison, 2022), which could present additional vulnerability during an environmental catastrophe. They highlight that there are no legal protections for sexual and gender minorities in the context of disasters under federal administrations that do not recognize them and in states that enact anti-LGBTQ+ policies.

In addition to the special risks that queer people face as a result of climate change and other environmental impacts, Whitley and Bowers (2023) showed that participants identifying as LGBT+ identified climate change as a greater threat, and endorsed more worry about climate change than cishet participants. This finding persisted even after the study authors controlled for other demographic factors that have been linked to climate change beliefs, including political party affiliation, ideology (conservative vs. liberal), white versus non-white racial identity and age (although worry about climate change only approached significance in the white, male, conservative subgroup). As a whole LGBT+ identity stood out in a way that led the authors to believe that the queer community is positioned to serve as a strong sect of climate activism.

LGBTQIA+ Activism

The LGBTQIA+ community presents with several strengths that predispose them to being powerful forces in the climate movement due to their unique political culture. They have shown an impressive ability to unite to fight for civil rights, and thus environmental groups have much to learn from them when it comes to uniting for the sake of a common cause.

To better understand strategies that sexual and gender minorities have used to respond to injustice, including coalition building, I met with Leo Goldsmith, a climate health specialist and doctoral student studying the intersection of LGBTQIA+ and environmental justice.

I met Goldsmith over Zoom and was immediately struck by his warm and open demeanor, as well as his humility. Goldsmith is in his early 30s and is in his second year of a PhD program at Yale School of Environment. Over the past six years, he has co-authored at least eight publications (most of which he is the first author), in addition to a long list of professional presentations, honors and awards. In addition, he serves as a board member for Out for Sustainability, an environmental advocacy group by and for LGBTQIA+ individuals.

Goldsmith (personal communication, February 25, 2025) grew up in financial hardship and started his career in food justice activism, which led him to learn more about environmental justice, including how social factors like race and class affect representation in climate policies and receiving resources. Goldsmith came out as transgender and queer in his young adulthood and saw the LGBTQIA+ movement as being tied to his sense of purpose as well. He sought to better understand how marginalization of the LGBTQIA+ community intersected with the marginalization of economically disadvantaged individuals and people of color in the context of environmental justice.

Goldsmith wants to help others understand how anti-LGBTQ+ policies create barriers to resources and services that these communities need for preparedness, response and relief in the event of a disaster (Communication with L. Goldsmith, February 25, 2025). According to Goldsmith:

> Living in states with anti-LGBTQ+ policies increases the chance of hate crimes and prejudiced events, which also increase during times of disasters and emergencies especially because some religious leaders and politicians will blame LGBTQ+ individuals for the disaster or emergency itself. We see that every single time a major disaster occurs. We saw that during Hurricane Sandy, Hurricane Harvey, and even the more recent wildfires in LA. But one thing I wonder and hope to do research on in the future, is if LGBTQIA+ people who are displaced from disasters take that as an opportunity to move to more affirming states.

In addition to advocating for queer rights in the context of environmental toxins and natural disaster response, another critical aspect of Goldsmith's platform is making the environmental justice space more inclusive to LGBTQ+ people (Goldsmith, personal communication, February 25, 2025). He referenced a 2022 dissertation by Jeff Feng, who spoke about the LGBTQ+ groups Queer Bloc being pushed to the back of the NYC Climate March of 2014 (Feng, 2022).

Goldsmith acknowledged that climate awareness among the LGBTQ+ community itself is still emerging. This is partly because climate change does not impact on all groups equally. At the same time, part of Goldsmith's mission is to debunk notions that LGBTQ+ individuals are mostly white and wealthy and to help the public understand that intersecting minoritized identities render them at greater risk of climate impacts as well.

Goldsmith also discussed organizations that are taking on both the environmental justice and LGBTQIA+ justice space, such as Out for Sustainability and Queers 4 Climate Justice, as well as efforts of established environmental organizations, such as the Sierra Club to release solidarity statements during Pride months. According to Goldsmith, "Grassroots groups like *Espicy Nipples* in Puerto Rico that are also acutely aware of the impacts of disasters and how homophobia, transphobia, sexism, racism, colonialism, ableism and more are connected" (personal communication, February 2025).

In addition to gaining Leo Goldsmith's insights about how queer people are affected by the climate crisis, I also wanted to gain knowledge of the LGBTQIA+ community's strategies for solidarity. I was interested in learning about aspects of queer culture that have enabled constituents of the LGBTQIA+ movement to overcome barriers to unification.

When asked to share his perspective on the queer community's strategies for unification, Goldsmith emphasized the importance of art, creativity, and joy. He explained that people have dismissed queer activists, misunderstanding them as being childish. In reality, expressions of joy from drag, to dance parties, to the bright and colorful Pride flag, are intentional and strategic. According to Goldsmith, "Joy is a childish thing, but those are things that synergize with one another." He also stated, "How we get through these tough times [is through] exercises of imagination which I think is also very important for climate anxiety and eco-anxiety."

Goldsmith and I also spoke about an important structure within the queer community that serves as a foundation for solidarity: The chosen family. Part of the reason why poverty and social isolation are problems faced by LGBTQIA+ elders is that they came of age at a time when gay marriage was illegal. They suffered trauma from losing friends and romantic partners to AIDS, and some were hesitant to be intimate with a partner out of fear of losing another loved one to this devastating illness. Many of these individuals did not come out to their families or were abandoned by their families. The chosen family is thus one that is not biological but still serves many of the same functions as biological families.

Chosen families can have the same protective effects as supportive biological ones (communication with Leo Goldsmith, February 25, 2025).

Sometimes, a member of the chosen family serves as the mother figure, who takes on a caretaker role. While chosen families are just as important as biological ones in mitigating the effect of traumatic events like environmental disasters, they are not recognized by society and therefore will not be taken into consideration in emergency management and recovery. For example, members of a chosen family may be separated in a shelter because they are not seen as a "real family." Goldsmith also explained that queer individuals can have alternate relationships that non-LGBTQIA+ culture doesn't recognize as valid. For example, for an aromantic person, a platonic relationship can bring as much intimacy as a romantic relationship does for others.

Goldsmith (personal communication, February 25, 2025) stressed that coalition building is an essential part of the LGBTQIA+ movement's strength. He shared a story, written by Jason Daur about an unlikely bond that emerged between queer individuals and conservative, rural farmers in the Philippines (Daur, 2021). The farmers had organized to protest the government taking their land. When students helped the farmers by taking care of weeds and raising funds for farm equipment, they nurtured a relationship with the farmers and the farmers wanted to learn more about LGB. Ultimately, they participated in Manila's Pride parade. Daur explains that they overcame their differences by uniting in the common cause of justice.

In this example, the heart of coalition building is similar to the concepts explored by Mónica Guzmán (2022) and similar to the example I described above about working with anti-vax parents. The idea here is that building a relationship, and ultimately an alliance, is prioritized over differences in beliefs. The other concept that is important to revisit here is one from Chapter 2 about differences in time perception, specifically the idea of diving in even if you don't know for sure that the outcomes will be favorable. You help to help without knowing for sure if the other will embrace, or even accept who you are as a result of you helping them. It's hard to invest time and energy into relationships that might not give back, but the chance they will can be too compelling to not try.

Ecofeminism – How Empowering Women Will Help Save the Earth

The intersection of climate change and the subjugation of women is portrayed with poetic, heart-wrenching detail in the 2017 film *Mother!* directed and written by Darren Aronofsky (2017). In *Mother!* Jennifer Lawrence (Mother) plays the romantic partner of a famous poet living in the middle of a valley with no other signs of civilization in sight. Civilization finds them, however, when they are bombarded by intrusive guests who destroy their home and ignore Mother's pleas to stop. Although it is not explicitly stated in the film, it is apparent that Lawrence's character represents Mother Earth, and her partner, "Him" (Javier Bardem) is God. While Him is portrayed as generous, welcoming, and forgiving of his catastrophic guests, in contrast, he is abusive toward Mother in enabling the guests. The film is segmented into several forms of destruction (e.g. flood) that Mother must repair but without all of humanity first being destroyed.

Our productivity and profit-obsessed society has dismissed the roles often designated to women. Roles of mother and nurturer are disregarded in favor of traditional masculine enterprises of capitalism and material wealth. This parallels the destructive forces that got us into our global mess.

On one hand, while women have made great advances to be included in the workplace, they are still underrepresented in leadership positions. On the other side of the coin, society has historically refused to recognize the value of those parents who choose to stay home to raise their children, the majority of whom are women. Allowing people more spiritual nourishment through time with their families (whether this be with their children, spouse, parents, or pets) would hypothetically lessen our incessant craving for acquiring material items, a habit that is literally destroying our habit and future chances for species survival.

From the perspective of biological sex, females are more vulnerable to the effects of toxins given that they have a higher proportion of bodily fat (Le Magueresse-Battistoni, 2020). Environmental pollutants have been hypothesized to adversely affect neurodevelopment (Parrón-Carrillo et al., 2024), thereby placing a heavier burden on the already serious responsibility felt by pregnant women to keep their unborn child safe and healthy.

As explained by environmental activist Joanna Macy and Chris Johnstone (2022), human history has been dominated by a power model where one has power over another. In order to attain this level of power, one needs to push others out. Having power does not guarantee security, as you have to contend with the jealousy of those below you, who will grab any chance to supplant you. Therefore, our current model of power is one that engenders fear, rigidity, and being closed off to any notion that you might be wrong.

Macy and Johnstone (2022) advocate for a cultural shift away from *power over* to *power with*. This entails several shifts, including pivoting from personal to team success, and from being motivated by the prospect of external reward (e.g. money and recognition) to the satisfaction of understanding and responding to your inner calling. It requires understanding that while on the surface, your efforts toward a cause might seem inconsequential, by playing your part you will add up to small ripples, that together create bigger waves.

Of course, not everyone has the privilege to have the resources and energy to seek their inner calling. Many of earth's inhabitants struggle to feed their families and keep a roof over their heads. With that, we now turn to another aspect of social identity that is of great relevance to the climate crisis: Socio-economic status.

SOCIO-ECONOMIC STATUS

Ultimately, climate change will affect us all. No amount of money can renew a depleted and unsustainable earth. However, the fact that climate change has already started to hurt and kill humans is glaringly obvious when we look at the effect of natural disasters on those who are economically disadvantaged.

Socio-economic status (SES) is a variable that encompasses several factors, including individual and community economic resources, as well as educational attainment and social status, both of which are often tied to wealth.

Countless natural disasters have disproportionately affected economically disadvantaged people. The May 24, 2024 landslide in Papua New Guinea, which buried alive thousands of villagers, is believed to have been caused by a combination of record heavy rains and tectonic shifts that caused an earthquake (Li et al., 2024). The land had been overused for mining for years. The prime minister of New Guinea publicly linked climate change to the disaster (Jose & Needham, 2024). The aftermath of the 2004 Tsunami in Asia, which killed about 200,000 people, revealed the exacerbated risk of natural disasters for economically disadvantaged women and children. In this case, most of the residents who were in their homes on the Sunday morning it struck were women and children, while the men typically ran errands (MacDonald, 2005). Women and children, particularly those who are disadvantaged, are less likely to have access to opportunities to learn how to swim.

Natural disasters in the United States have also exposed how income disparity renders poorer communities more susceptible to exposure to environmental toxins. The Flint water crisis of 2014 (CDC, 2024) is a case in point. In order to save money, the town of Flint, Michigan changed its municipal water supply to the Flint River which was polluted with toxins. The government turned a deaf ear toward people pleading for help, some of whom came carrying foul-looking water, and many of whom could not afford bottled water. Action was taken only after it was discovered that children living in the area tested high for lead, a neurotoxin that can have persistent and/or permanent cognitive effects on children (Schneider, 2023).

People living in poverty are more likely to live in areas where they are exposed to higher levels of environmental toxins, such as pesticides (Donley et al., 2022). In addition, the World Economic Forum (World Economic

Forum, 2023) highlights the following vulnerability factors for the economically disadvantaged in the face of climate change:

1. Health via increased incidence of mosquito-borne pathogens such as malaria, increased water shortages, and resulting diarrheal deaths from contaminated water, and limited access to health care.
2. Hunger via changes in food production and prices.
3. Water scarcity.
4. Education will be compromised as more children will need to work to compensate for their families' economic stress.
5. Creating refugees – climate migration has already begun and is anticipated to worsen in the coming years, with an estimation that by 2050, 216 million climate refugees will have been displaced in areas such as sub-Saharan Africa, East Asia and The Pacific, and South Asia.
6. Work-related hazards – excessive heat will create more missed days of work and illness due to heat.

One misconception is that concern for climate change is reserved for the privileged. Research supports the notion that higher SES individuals tend to show more consistency between climate change beliefs and pro-environmental behavior, but remember that SES is not only about income, but also level of education. According to a study by Kim et al. (2024) education level moderated consistency between climate change beliefs and pro-environmental behavior, while income did not.

The researchers' (Kim et al., 2024) choice of volunteer time as an operational definition, or way of defining and measuring pro-environmental behavior, is worth attention. Individuals with greater wealth might have more discretionary time to volunteer.

The relevance of educational level and the idea of time wealth also comes into play when examining the implications of creating green jobs. The idea of creating more green jobs came up at a board of education meeting for my children's previous district two years ago. The district was offering internship opportunities for high school students interested in careers in wind energy. The administrator mentioned that it was meant for students who are strong in math and science. My ears perked up at this mention of qualifications. We need to create learning and work opportunities that are accessible to individuals across the spectrum of access to educational opportunities and skill level. We need to help all individuals feel that they are valuable and that all are welcome and needed to engage in this important work.

As we will discuss in Chapter 5, conservation is ingrained in our psyches in different ways and allowing people to express creativity in a resourceful way can empower them with a sense of purpose.

RACIAL, ETHNIC, AND CULTURAL IDENTITY

The Diagnostic and Statistical Manual of Mental Health Disorders (DSM-V; American Psychiatric Association, 2013) includes the Cultural Formulation Interview (CFI) – a series of questions that guide clinicians in getting a better understanding of the cultural, racial, and ethnic backgrounds of clients and importantly, how these aspects of identity factor into a case conceptualization. A case conceptualization is like a roadmap that psychologists use to guide treatment. Later in this book, I describe in greater depth what a CBT case conceptualization for eco-distress might look like.

For now, I emphasize that the point of a cultural component in a case conceptualization is not to suggest that there is one treatment approach that is good for, say an Asian American client and another that is better for a Black Caribbean one. Not only would this not be helpful, but it would fuel stereotypes and therefore be harmful to the client.

Rather, incorporating a cultural component in case formulation serves the following purposes: One, it acknowledges that clients from different cultural, racial, and ethnic backgrounds see the world through the lens of their culture and by understanding how they experience their problems through this lens, a clinician will be more effective at helping the client. Research supports the notion that clinicians who demonstrate stronger cultural humility, or the ability to self-reflect on how one's own biases are affecting the treatment, provide treatment that has better outcomes (Zhang et al., 2022).

Another purpose behind incorporating a cultural component is that it acknowledges the likely possibility that a minoritized client has been discriminated against to some extent, whether this be in the form of a covert microaggression or a more blatant injustice. Understanding the client through the lens of social injustice allows the clinician to avoid inaccurately jumping to the conclusion that a client's perceived discrimination is due to irrational thinking.

One approach recommended in cultural competence training (Calloway, 2024) is to ask the client to consider what the discrimination they experience means about them as a person. The goal then becomes helping the client avoid personalizing discrimination, so as to avoid the pitfalls of a victim mentality, while also validating and affirming the pain of being the target of discrimination.

In addition to respecting cultural values and, when necessary, adapting clinical practices to show greater sensitivity to them, it is essential for a culturally sensitive clinician to appreciate the societal marginalization faced by people of color. This is a fact that often intersects with wealth disparity, with wealth gaps being particularly prominent among Black Americans and Latine Americans (Kochhar & Moslimani, 2023).

In contrast to the myth that climate change is a concern of the privileged only, Sasser and Merchant's research (2024) showed that nonwhite respondents to a survey assessing participants' thoughts about birthing and raising children in the context of climate change were more likely than white respondents to report planning on having fewer future children than they wanted due to climate-related emotions.

While we should all advocate for minoritized and marginalized communities, history has shown that the strongest advocates are individuals who represent affected groups. Nelson Mandella, Martin Luther King Jr., and Sojourner Truth are just a few of the many examples. Therefore, climate change research should include the perspectives of Black, Indigenous, and People of Color (BIPOC) communities, and these perspectives should be incorporated into community disaster response and sustainability planning.

Research has begun to address this need, and studies indicate that people of color express more concern about environmental risks associated with climate change across the political spectrum (Ballew et al., 2021). The study found less polarization in African Americans and Latine in that racial/ethnic differences were only partially accounted for by ideological identification or political party affiliation. Across the political spectrum, people of color expressed more concern about how climate change will endanger them personally, or in other words, they were more aware of the risks posed by climate change.

In addition to encouraging leadership in environmental policies among people of color, we cannot discuss skills for unification without addressing the tensions that continue to exist among different cultural and racial groups. For some time, I wondered how Radically Open Dialectical Behavior Therapy could address these tensions, as this intervention is aimed at helping people feel more socially connected. Luckily, the RO DBT community has had active discussions about skills for empowering BIPOC individuals.

In her blog post, *How I am Unlocking Freedom: RO DBT Skills for BIPOC Navigating Social Signals and Racism* (2024), Michelli Simpson validates the need for BIPOC individuals to be vigilant given the reality of systemic racism and her personal history of being the victim of hostility. She shares how she coped with these realities by "putting on a mask of agreeability" (Simpson, 2024, para. 2) in order to ensure that others would accept her. While this has been necessary at times, she also recognized that, when used excessively, it came with the unfortunate side effects of mental exhaustion and not being able to express her authentic self.

Simpson (2024) highlights how she learned to respond to protective hyperawareness with self-compassion given that it is a survival tactic, while at the same time practicing questioning assumptions that people are mistreating her. She recommends fostering curiosity, such as wondering if someone who seems standoffish might simply be having a bad day; openly expressing her emotions, rather than withdrawing in response to a microaggression (the example she gives is, "Ouch, that felt a bit hurtful"), and using non-verbal communication such as a closed-mouth smile and an open, relaxed posture, in order to signal approachability and warmth.

At the same time, it is important for therapists to show cultural competence, or skills, for using language and interviewing techniques that communicate a spirit of diversity (e.g. Calloway, 2024).

The way one person sees reality can be completely different from the next. Only by celebrating diversity can we transcend the notion that there is one right way to solve climate change, or the notion that playing a small role in the solution is worthless to the extent that it is equivalent to not doing anything at all. Indeed, great steps will need to be taken to adequately address climate change, but this will be best accomplished by the actions of many weaving together into a more impactful whole, and we can't get there if we are unnecessarily judgmental and critical in our advocacy.

Culturally sensitive mental health treatment for the eco-distressed should also consider cultural traditions that involve conservation and sustainability and strive to honor these traditions. Indigenous communities (as discussed further in Chapter 5) offer excellent examples of cultural practices that involve sustainability (e.g. Recio & Hestad, 2022). A clinician working with an eco-distressed client can ask the client to reflect on sustainability that is already baked into their cultural traditions.

ENVIRONMENTAL IDENTITY

Environmental Identity (Clayton, 2003; 2021) is a facet of self-identity that refers to how one sees oneself in relation to the natural world. Susan Clayton, a psychology professor who has written extensively about the psychology of conservation and sustainability, developed the Environmental Identity Scale in order to assess the following facets of environmental identity: (1) The extent of salience and meaning of one's relationship with nature; (2) one's perception of how nature is connected with a community with which one identifies; (3) a shared value system of sustainability and conservation; (4) positive emotions experienced through the community one identifies with that relates to deriving pleasure through nature, such as appreciating it aesthetically; and (5) seeing nature as part of your life story. Example items include "I like to spend time outdoors in natural settings (such as woods, mountains, rivers, fields, local parks, lake or beach, or a leafy yard or garden);" "I think of myself as a part of nature, not separate from it;" and "I can consider myself a steward of the natural environment."

The Environmental Identity Scale (Clayton, 2003) can be used as a tool for better understanding clients presenting with eco-distress as well as to alert clinicians that environmental concerns might influence their beliefs about themselves and others. Clinical psychologist Thomas Doherty has integrated the concept of environmental identity into his work with eco-distressed clients (Doherty et al., 2024). Some examples of exercises he uses include: (1) Creating an environmental family tree to better understand how you and your family have related to nature. (2) Creating a timeline of your life story in relation to nature; and understanding how one expresses their environmental values (e.g. as a human concern given the impacts to humanity of climate change, versus a love for wild life at risk, vs. solastalgia, or the angst experienced in response to loss of the aesthetic pleasure derived from nature in response to environmental devastation). These approaches can be integrated with other aspects of the client's identity to help the whole person as opposed to hyper-fixating on the "problem" such as eco-distress or activist burnout.

CONCLUSION: UNDERSTANDING IDENTITIES IS CRITICAL TO SOLIDARITY

Identity has a critical place in understanding human motivation, attitudes, and behavior. We need to appreciate human diversity and complexity in order to mobilize our emotional reactions toward innovative solutions. In this chapter, I presented two perspectives that are not mutually exclusive, but rather, have the potential to create a powerful synergy: The idea, on the one hand, of identifying with the cause of environmentalism, and, on the other, the idea of funneling environmentalism through the lens of a long-standing, established facet of one's identity, including gender, cultural, racial, personality, etc. Finally, I have made the case for the idea that an understanding of one's identity must be paired with an appreciation of the diversity of other identities to feel empowered to address eco-distress in a meaningful way.

REFERENCES

American Psychiatric Association. (2013). *Diagnostic and statistical manual of mental disorders* (5th ed.). https://doi.org/10.1176/appi.books.9780890425596

Aronofsky, D. (Director). (2017). *Mother!* [film]. Paramount Pictures.

Ballew, M. T., Pearson, A. R., Schuldt, J. P., Kotcher, J. E., Maibach, E. W., Rosenthal, S. A., & Leiserowitz, A. (2021). Is the political divide on climate change narrower for people of color? Evidence from a decade of U.S. polling. *Journal of Environmental Psychology, 77,* 101680. https://doi.org/10.1016/j.jenvp.2021.101680

British Broadcast Company (BBC). (2024, May 9). *Greta Thunberg: Who is the climate activist and what has she achieved?* https://www.bbc.com/news/world-europe-49918719

Brooks, D. (2024, September 4). How Trump wins (and Harris and the Democrats blow it. *The New York Times.* https://www.nytimes.com/2024/09/04/opinion/trump-win-election-harris.html?smid=nytcore-ios-share&referringSource=articleShare&sgrp=c-cb

Brown, A. (2023, June 23). *5 key findings about LGBTQ+ Americans.* Pew Research Center. https://www.pewresearch.org/short-reads/2023/06/23/5-key-findings-about-lgbtq-americans/

Cain, C. M., & Ellison, J. M. (2022). *The pains of imprisonment through a rainbow lens: An overview of the marginalized conditions of incarceration for LGBTQ persons.* Routledge.

Calloway, A. (2024, July 26). *Delivering culturally responsive cognitive behavioral therapy* [Webinar]. Association for Behavioral and Cognitive Therapies.

Carter, B. (2021). *Impact of social inequalities and discrimination on vulnerability to crises.* K4D Helpdesk Report 994. Institute of Development Studies. https://doi.org/10.19088/K4D.2021.049

Centers for Disease Control and Prevention (CDC). (2024, April 8). *Story: Flint Water crisis.* CASPER. https://www.cdc.gov/casper/php/publications-links/flint-water-crisis.html#:~:text=Background,declared%20on%20January%2016%2C%202016

Cipriani, E., Frmento, S., Gemignani, A., & Menicucci (2024). Personality traits and climate change denial, concern, and proactivity: A systematic review and meta-analysis. *Journal of Environmental Psychology, 95,* 102277. https://doi.org/10.1016/j.jenvp.2024.102277

Clague, C. A., & Wong, Q. (2023). Social anxiety and its maintaining factors: Accounting for the role of neuroticism. *Journal of Psychopathology and Behavioral Assessment, 45,* 469–479. https://doi.org/10.1007/s10862-023-10030-2

Clayton, S., Czellar, S., Nartova-Bochaver, S., Skibins, J., Salazar, G., Tseng, Y.-C., Irkhin, B., & Monge-Rodriguez, F. S.(2021). Cross-cultural validation of a revised environmental identity scale. *Sustainability, 13,* 2387. https://doi.org/10.3390/su13042387

Clayton, S. D. (2003). Environmental identity: A conceptual and an operational definition. In S. D. Clayton & S. Opotow (Eds.), Identity and the natural significance of nature (pp. 45–65). The MIT Press.

Corbett, S. (2016, November). *Activism needs introverts* [Video]. TED Conferences. https://www.ted.com/talks/sarah_corbett_activism_needs_introverts?language=en

Corbett, S., & Housley, S. (2011). The craftivist collective guide to craftivism. *Utopian Studies, 22*(2), 244–351. https://doi.org/10.5325/utopianstudies.22.2.0344

Craftivist Collective. (n.d.) *Our story.* Retrieved April 10, 2026, from https://www.craftivist-collective.com/our-story

Curran, T., & Hill, A. P. (2019). Perfectionism is increasing over time: A meta-analysis of birth cohort differences from 1989 to 2016. *Psychological Bulletin, 145*(4), 410–429. https://doi.org/10.1037/bul0000138

Dalio, R., Grant, A., Little, B., & Golden, J. (n.d.). *Research & development of our assessment.* Principles You. www.principlesyou.ccom

Daur, J. (2021, May 4). *A perspective on queer solidarity.* Unbound: The Intersections of Faith and Justice. https://justiceunbound.org/a-perspective-on-queer-solidarity/

De Jong, N., Wisse, B., Heesink, J. A. M., & van der Zee, K. I. (2019). Personality traits and career role enactment: Career role preferences as a mediator. *Frontiers in Psychology, 10,* 1720. https://doi.org/10.3389/fpsyg.2019.01720

Desilver, D. (2022, March 10). *The polarization in today's congress has roots that go back decades.* Pew Research Center. https://www.pewresearch.org/short-reads/2022/03/10/the-polarization-in-todays-congress-has-roots-that-go-back-decades/

Doherty, T., Artman, S., Homan, J., Keluskar, J., & White, K. (2024). Environmental identity-based therapies for climate distress: Applying cognitive behavioural approaches. *The Cognitive Behaviour Therapist, 17,* e29, 1–16. https://doi.org/10.1017/S1754470X24000278

Donley, N., Bullard, R. D., Economos, J., Figueroa, I., Lee, J., Liebman, A. K., Martinez, D. N., & Shafiei, F. (2022). Pesticides and environmental injustice in the USA: Root causes, current regulatory reinforcement and a path forward. *BMCC Public Health, 22*(708), 1–23. https://doi.org/10.1186/s12889-022-13057-4

Federico, C. M., & Malka, A. (2018). The contingent, contextual nature of the relationship between needs for security and certainty and political preferences: Evidence and implications. *Political Psychology, 39*(S1), 3–48.

Federico, C. M., & Malka, A. (2023). The psychological and social foundations of ideological beliefs systems. In L. Huddy, D. O. Sears, J. Levy, & J. Jerit (Eds.), *The Oxford handbook of political psychology* (3rd ed.). Oxford UP. https://psycnet.apa.org/doi/10.1093/oxfordhb/9780197541302.013.16

Feng, J. (2022). *Toward queer climate justice.* UC Santa Barbara Electronic Theses and Dissertations. Peer Reviewed Thesis/Dissertation. Scholarship.org

Flanagan, C. A., & Tucker, C. J. (1999). Adolescents' explanations for political issues: Concordance with their views of self and society. *Developmental Psychology, 35*(5), 1198–1209. https://doi.org/10.1037/0012-1649.35.5.1198

Goldsen, K. (2018). 'Shifting social context in the lives of LGBTQ older adults'. *Public Policy and Aging Report, 28*(1), 24–28. https://doi.org/10.1093/ppar/pry003

Goldsmith, L., & Bell, M. (2022). Queering environmental justice: Unequal environmental health burden on the LGBTQ+ community. *New Frontiers of Environmental Justice, a Journal of the American Journal of Public Health, 112*(1), 79–87. https://doi.org/10.2105/AJPH.2021.306406

Goldsmith, L., Raditz, V., & Méndez, M. (2022). Queer and present danger: Understanding the disparate impacts of disasters on LGBTQ+ communities. *Disasters, 46*(4), 946–973. https://doi.org/10.1111/disa.12509

Gray, L., McNeill, B., Pecora, L., Macfarlane, S., Hayley, A., Hitch, D., & Evans, S. (2025). Navigating neurodivergence: A scoping review to guide health professions educators. *Medical Education.* 1–12. https://doi.org/10.1111/medu.15676

Guzmán, M. (2022). *I never thought of it that way: How to have fearlessly curious conversations in dangerously divided times.* BenBella Books.

Healy, J. (2025a, May 28). A Missouri town was solidly behind Trum. Then Carol was detained. *The New York Times.* https://www.nytimes.com/2025/05/28/us/missouri-immigrant-trump.html?searchResultPosition=1

Healy, J. (2025b, June 4). 'Carol,' whose detention rattled her small Missouri town, is released. *The New York Times.* https://www.nytimes.com/2025/06/04/us/politics/carol-missouri-migrant.html

Jose, R., & Needham, K. (2024, May 29). Papua New Guinea leader Marape says extraordinary weather causes disasters. *Reuters.* https://www.reuters.com/world/

asia-pacific/papua-new-guinea-leader-marape-says-extraordinary-weather-causing-disasters-2024-05-29/#:~:text=SYDNEY%2C%20May%2029%20(Reuters), which%20may%20have%20killed%20thousands.

Keluskar, J. (2024, January 5). Dealing with disgust. *Psychology Today*. https://www.psychologytoday.com/us/blog/alone-together/202401/dealing-with-disgust

Kim, H. S., Eom, K., Panzone, L. A., & Sherman, D. K. (2024). Why do I act for the environment? Socioeconomic status moderates the relationship between climate change beliefs and sustainable actions. *Motivation Science*. Advance online publication. https://doi.org/10.1037/mot0000343

Kochhar, R., & Moslimani (2023, December 4). *Wealth surged in the pandemic, but debt endures for poorer black and Hispanic families*. Pew Research Center. https://coilink.org/20.500.12592/d51c897

Lawson, D. F., Stevenson, K. T., Peterson, M. N., Carrier, S. J., Strnad, R. L., & Seekamp, E. (2019). Children can foster climate change concern among their parents. *Nature Climate Change, 9*, 458–462. https://doi.org/10.1038/s41558-019-0463-3

Le Magueresse-Battistoni, B. (2020). Adipose tissue and endocrine-disrupting chemicals: Does sex matter? *Environmental Research and Public Health, 17*(24), 9403. https://doi.org/10.3390/ijerph17249403

Li, Z., Li, W., Xu, Q., Pu, F., Yu, W., Shan, Y., Guo, P., Yu, Zhou, S., Pu, C., & Wang, X. (2025). Brief report on the catastrophic landslide in Papua New Guinea on May 24, 2024. *Landslides, 22*, 1877–1889. https://doi.org/10.1007/s10346-025-02511-0

Linden, D. J. (2020). *Unique: The new science of human individuality*. Basic Books.

Lionetti, F., Aron, A., Aron, E. N., Burns, G. L., Jagiellowicz, J., & Pluess, M. (2018). Dandelions, tulips and orchids: Evidence for the existence of low-sensitive, medium-sensitive and high-sensitive individuals. *Translational Psychiatry, 8*(24), 1–11. https://doi.org/10.1038/s41398-017-0090-6

MacDonald, R. (2005). How women were affected by the Tsunami: A perspective from Oxfam. *Plos Medicine*. https://doi.org/10.1371/journal.pmed.0020178

Macy, J., & Johnstone, C. (2022). *Active Hope: How to face the mess we're in with unexpected resilience & creative power* (revised ed.). New World Library.

Mazefsky, C. A., & White, S. W. (2014). Emotion regulation: Concepts & practice in autism spectrum disorder. *Child and Adolescent Psychiatric Clinics of North America, 23*(1), 15–24. https://doi.org/10.1016/j.chc.2013.07.002

McKenney, E. E., Richards, J. K., Day, T. C., Brunwassar, S. M., Cucchiara, C. L., Kofner, B., McDonald, R. G., Lynch, K. G., Lamm, J., Kang, E., Lerner, M. D., & Gotham, K. O. (2024). Satisfaction with social connectedness is associated with depression and anxiety symptoms in neurodiverse first-semester college students. *Autism: The International Journal of Research and Practice, 28*(8). https://doi.org/10.1177/13623613231216879

Milad, E., & Bogg, T. (2020). Personality traits, coping, health-related behaviors, and cumulative physiological health in a national sample: 10 year prospective effects of conscientiousness via perceptions of activity on allostatic load. *Annals of Behavioral Medicine, 54*(11), 880–892. https://doi.org/10.1093/abm/kaaa024

Moss, J. T., & O'Connor, P. (2020). Political correctness and the alt-right: The development of extreme political attitudes. *Plos One, 15*(10), e0239259. https://doi.org/10.1371/journal.pone.0239259

Nevid, J. (2022). *Essentials of psychology: Concepts and applications* (6th ed.). Cengage.

Parrón-Carrillo, R., Nievas-Soriano, B. J., Parrón-Carreño, T., Lozano-Paniagua, D., & Trigueros, R. (2024). Environmental exposure to pesticides and the risk of

child neurodevelopmental disorders. *Medicina, 60*(3), 475. https://doi.org/10.3390/medicina60030475
Pickering, G. J., & Dale, G. (2023). Trait anxiety predicts pro-environmental values and climate change action. *Personality and Individual Differences, 205,* 112101. https://doi.org/10.1016/j.paid.2023.112101
Pickering, G. J., & Dale, G. (2024). The role of perceived powerlessness and other barriers to climate action. *Canadian Geographies, 68,* 560–573. https://doi.org/10.1111/cag.12938
Pledl, C. (2021). Eco-ableism in the environmental justice movement. *Vermont Journal of Environmental Law, 23*(1), 1–26.
Recio, E., & Hestad, D. (2022). Indigenous peoples: Defending an environment for all. Policy Brief #36. Earth negotiations bulletin. *International Institute for Sustainable Development,* 1–10. https://www.iisd.org/system/files/2022-04/still-one-earth-Indigenous-Peoples.pdf
Roure, J. G. (2019). The reemergence of barriers during crises & natural disasters: Gender-based violence spikes among women & LGBTQ+ persons during confinement. *Whitehead Journal of Diplomacy and International Relations, 21,* 23.
Sasser, J. S., & Merchant, E. K. (2024). Climate emotions, parenting plans and racial difference in the United States. *The Journal of Climate Change and Health, 19,* 100346. https://doi.org/10.1016/j.joclim.2024.100346
Schleck, G., & Ben-Alon, L. (2024). Eco-ableism and access circularity in natural building. *Frontiers of Architectural Research, 13*(2), 235–248. https://doi.org/10.1016/j.foar.2023.11.005
Schneider, J. S. (2023). Neurotoxicity and outcomes from developmental lead exposure: Persistent or permanent? *Environmental Health Perspectives, 131*(8): 085002. https://doi.org/10.1289/EHP12371
Sears, B. (2024). *Housing is an LGBT issue.* UCLA Other Recent Work. Scholarship.org. https://escholarship.org/uc/item/7r35n1tb
Setti, A., Lionetti, F., Kagari, R. L., Motherway, L., & Pleuss, M. (2022). The temperament trait of environmental sensitivity is associated with connectedness to nature and affinity to animals. *Heliyon, 8*(7), e09861. https://doi.org/10.1016/j.heliyon.2022.e09861
Shin, K. E., & Newman, M. G. (2019). Self- and other-perceptions of interpersonal problems: Effects of generalized anxiety, social anxiety, and depression. *Journal of Anxiety Disorders, 65,* 1–10. https://doi.org/10.1016/j.janxdis.2019.04.005
Simpson, M. (2024, March 28). *How I am unlocking freedom: RO DBT skills for BIPOC Navigating social signals and racism* [Blog post]. Radically Open Blended Learning. https://www.radicallyopen.net/blog/
Stokes, R. G., & Stokes, M. A. (2018). Gender identity and sexual orientation in autism spectrum disorder. *Autism, 22*(8), 970–982.
Taylor, E. C., Livingston, L. A., Callan, M. J., Hanel, P. H. P., & Shah, P. (2021). Do autistic traits predict pro-environmental attitudes and behaviors and climate change belief? *Journal of Environmental Psychology, 76,* 101648. https://doi.org/10.1016/j.jenvp.2021.101648
The Trevor Project. (2021). *National survey on LGBTQ youth mental health 2021.* https://www.thetrevorproject.org/survey-2021/?section=Introduction
The Williams Institute, UCLA School of Law. (2021, January). *LGBT demographic data interactive.* Los Angeles.
Thunberg, G. (2023). *The climate book: The facts and the solutions.* Penguin Press.
Watfern, C., & Carnemolla, P. (2025). Inclusive climate action: A scoping review on the representation and inclusion of people with intellectual disability in climate change research. *British Journal of Learning Disabilities, 53*(3), 1–18.

West, M., & Yockey, K. (2023, August 24). What to know about antidepressants and heat tolerance. *Medical News Today.* https://www.medicalnewstoday.com/articles/antidepressants-and-heat

White, H. (2022). *One green thing.* Harper Horizon.

Whitley, C. T., & Bowers, M. M. (2023). Queering climate change: Exploring the influence of LGBTQ+ identity on climate change belief and risk perceptions. *Sociological Inquiry, 93*(3), 413–439. https://doi.org/10.1111/soin.12534

World Economic Forum (2023, January 13). The climate crisis disproportionately Hits the poor. How can we protect them? Forum Institutional. *World Economic Forum.* https://www.weforum.org/stories/2023/01/climate-crisis-poor-davos2023/

Yoneda, T., Lozinski, T., Turiano, N., Booth, T., Graham, E. K., Mrozek, D., & Terrera, G. M. (2023). The big five personality traits and allostatic load in middle to older adulthood: A systematic review and meta-analysis. *Neuroscience and Biobehavioral Reviews, 148,* 105145. https://doi.org/10.1016/j.neubiorev.2023.105145

Young, M. (2021, December 27). How disgust explains everything. *The New York Times Magazine.* https://www.nytimes.com/2021/12/27/magazine/disgust-science.html

Zawadzka, A. M., Besta, T., & Iwanowska, M. (2024). Various faces of activist identity: The role of intrinsic and materialistic goals and mass gatherings. *Journal of Community & Applied Social Psychology, 34*(3), e2798.

Zhang, H., Watkins, C. E. Jr, Hook, J. N., Hodge, A. S., Davis, C. W., Norton, J., Wilcox, M. M., Davis, D. E., DeBlaere, C., & Owen, J. (2022). Cultural humility in psychotherapy and clinical supervision: A research review. *Counseling and Psychotherapy Research, 22*(3), 548–557. https://doi.org/10.1002/capr.12481

Chapter **4**

The Effect of World Views and Beliefs on Climate Responses

In a blog post I wrote for *Psychology Today* entitled "The Psychology of Climate Change" (Keluskar, 2023), I discuss how my reluctance to be explicit about my climate values led to pro-environmental efforts backfiring. I share a story, for instance, about when my son was a toddler, and I sent him to day care with a reusable plate. I later learned that his caretakers would first dole out his portion on a paper plate and then transfer the food to the reusable plate. They thought it was simply "Colby's special plate" and were unaware of my environmental intentions. To be fair, I hadn't stated my intentions because I was worried about being judged as a difficult parent.

I'm not alone in feeling awkward about bringing up climate change, as a 2022 study published in *Nature Communications* (Sparkman et al., 2022) showed that people underestimate how much others are concerned about climate change. Collectively, this represents an assumption we are making about others' views of the world.

Our views of self, the world (including our view of others), and our future (Beck et al., 1979 as cited by Persons, 2008) shape the contour of our daily experiences and form the root of who we are. They can also affect how we experience and respond to eco-distress, which ranges on a continuum from being concerned about climate change without this concern affecting well-being and relationships, to losing sleep and having trouble interacting with others. Understanding a client's beliefs in these three areas is a fundamental part of the cognitive behavioral psychologist's toolkit, serving as a thread that can stitch together painful emotions, thoughts, events, and behaviors reported by the client into a meaningful whole and guide the clinician in asking exploratory questions that lead the client to helpful insights.

While concern is an adaptive response to the climate crisis, our belief systems can exacerbate distress to the point where it is no longer motivating

DOI: 10.4324/9781003489665-5

proactive solution-seeking behavior, but rather, leading to less helpful behaviors like self-destructive ones or avoidance.

In reality, we are all hurting the environment to an extent, and this could lead to eco-distress becoming a form of what social scientists call *moral frustration*, the less intense variation of moral injury, which is psychological pain triggered by seeing yourself as someone who is acting unethically and in a way that is hurtful to others (Litz, 2022).

According to clinical psychologist and military trauma expert Brett Litz (2022), moral frustration is not a clinical problem, while moral injury causes functional impairment, such as problems with daily living tasks like work and sustaining relationships. Still, we can gain insights into addressing the moral frustration of being perpetrators of climate change from Litz' work with individuals struggling with moral injury. Litz encourages people to avail themselves of opportunities to do good for the world toward the goal of helping the client restore faith in personal and collective humanity.

According to Litz (2022), who has worked with veterans who grapple with moral injury for their actions in combat, it is challenging to sit with the existential dread of moral harm. In the approach he developed called *adaptive disclosure*, he helps clients respond adaptatively to the shame of their moral transgressions through actions like imagining having a conversation with the person you harmed or writing them a letter. This can be extended to eco-distress via an imagined correspondence with future generations who will inherit the earth. Therapists can guide their client in writing a letter that expresses their positive intentions while acknowledging the human limitations that held them back from doing more to stop environmental devastation.

Litz (2022) also recommends helping clients understand that their regretted actions don't define them as a person. Additionally, he recommends encouraging clients to take reparative actions. These actions could be as simple as showing kindness to others while you are out at a park, and what is even better is adding virtuous acts, all with the goal being to enhance compassion and a sense of shared humanity with others. This is not too different from advice given by climate-aware clinicians that encourage clients to engage in collective climate action.

Exploring a client's beliefs and how they manifest in their perceptions of the world, and consequently, their emotional well-being is at the heart of a cognitive behavioral therapist's work.

When it comes to health, well-being, and longevity, individuals who lean toward positive beliefs, such as in the self being worthy of love and the world as safe and just, fare better than those with negative beliefs, such as the world being a dangerous place (e.g. Clifton & Meindl, 2021).

How does this play out in the context of climate change where gloom and doom are more of a reality? This question dovetails with the broader issue of addressing to what extent eco-distress is adaptive. The short answer to that question is yes, to an extent, but we need to be clear and consistent as a profession when it comes to the role that psychotherapists should play in guiding a client toward healthy, pro-environmental responses and away from self-destructive ones.

We also need to consider the role of cultural beliefs, specifically the values espoused by collectivistic versus individualistic cultures, as well as the role played by concepts such as social exclusion, fear of missing out (FOMO), optimism and pessimism, self-efficacy, and justice sensitivity. We start with a macro view, so to speak, by diving into cultural beliefs that stress individualism and those that prioritize collectivism.

INDIVIDUALISM VERSUS COLLECTIVISM

Individualistic cultures emphasize personal characteristics, such as intelligence, as well as personal accomplishments, such as material wealth. In contrast, collectivistic cultures, while also prizing positive qualities and achievements, tend to prioritize social roles and obligations (Nevid, 2022). Individualistic cultures value standing out in a unique way, while collectivistic cultures emphasize interdependence, conforming, and deference to family, elders, and the community. Many cultures in Asia, Africa, and Central and South America are thought to be collectivistic, while individualistic cultures are typically associated with Western countries, including the United States, Canada, and much of Western Europe. There are pros and cons to each orientation, with collectivistic cultures encouraging cooperation but also squelching dissonance, and individualistic cultures encouraging assertiveness but fueling greed and materialism.

Research has shown that collectivism is positively related to environmental attitudes and intentions across several different cultures, including Latine (Naiman et al., 2023) and Sri Lankan (Samarasinghe, 2012) samples, presumably due to the theme of taking responsibility for your behavior in caring for the community that is ingrained in a collectivistic orientation (Chwialkowska et al., 2020).

One study (Naiman et al., 2023) that compared Latine and non-Latine adults, with the Latine sample consisting of individuals (both US and foreign born and white and non-white Latines) of Cuban, Mexican, Puerto Rican, and "Other Latine" descent found that among Latine participants, collectivistic values, but not familism, or prioritizing the needs of family over self, predicted intentions to engage in pro-environmental behaviors. Among non-Latine participants, ethnic identity exploration predicted pro-environmental values.

This study (Naiman et al., 2023) suggests the need to consider motivations for pro-environmental behavior within a cultural context and shows the limitations of making blanket statements about collectivism being pro-environmental, since prioritizing family values can also run counter to pro-environmental values if such values are not adopted by other family members. At the same time, the study also highlights the benefits of exploring one's ethnic identity in enhancing what social scientists refer to as *reflexivity* or the ability to self-reflect on how one's assumptions about the world affect how they view others.

Similarly, Samarasinghe (2012) found that collectivistic attitudes were associated with pro-environmental attitudes in a Sri Lankan sample, but pro-environmental attitudes were associated with green consumer choices only among unmarried and younger adults in their sample. The lack of consistency between environmental attitudes and green behaviors has been referred to as the "green gap" (Chwialkowska et al., 2020).

Collectivism is an important component of more eco-conscious attitudes, but not the whole story. Chwialkowska et al. (2020) theorize that culture indirectly influences pro-environmental behaviors but that it is important to also incorporate other factors that might motivate pro-environmental behavior, such as femininity and uncertainty avoidance, or the extent to which members of a society seek orderliness and predictability.

Another study (Eom et al., 2016) specifically looked at cultural drivers of environmental behaviors in European American versus Japanese participants. While concerns about sustainability better predicted support for environmentalism in European Americans, perception of norms for environmentally friendly behavior, or the perception that such behavior is widespread, better predicted pro-environmental behaviors in Japanese participants. Therefore, collectivism had a stronger pull toward environmentalism in Japanese individuals, known to be a collectivistic culture, than it did among European Americans, who are associated with an individualistic culture. It should be noted that the study did not directly assess collectivistic and individualistic orientation.

We can also consider cultural values in the therapy room and tie them to pro-active sustainability in response to eco-distress. For example, we can help someone who relates to an individualistic culture see a unique sense of purpose in climate advocacy, and frame pro-environmental behaviors as a trait that helps them stand out, while we can engage collectivistic clients by guiding them in sustainable behaviors that are consistent with cultural norms, such as conservation practices in Mexican culture, or in Japanese culture, the concept of wabi-sabi, a decorative style that values simplicity and sustainable consumption (Fallahnejad, 2025).

Regardless of the extent to which you associate with a collectivistic or individualistic cultural identity, striving to feel a sense of belonging is a universal, psychological need. This need stems from the tribal nature of our species, and it is also why many feel distressed when they perceive themselves as being excluded.

SOCIAL EXCLUSION

Figure 4.1 *Drawing by Skylar Keluskar.*

A few years ago, when I was searching for a climate advocacy group, I came across one in New York City for parents to join with their young children. I reached out to the organizer who was eager to have me join a nature walk on Governor's Island in New York City. Given the formidable traffic of the New York metropolitan area, it was a hike for my family, but we made it happen on a Sunday afternoon in June and found it to be a rewarding experience. We couldn't join all their outings, but I expressed enthusiasm for being part of the group and requested being kept abreast of their events in case we could make it out to the city again. At first, the organizer acquiesced, but after a few events that we regretfully couldn't make, I stopped receiving emails from her.

I never found out what happened. While logic would tell me that she got busy or moved on to other endeavors, the sting of rejection hit me automatically like a knee jerk response. I struggle with a phenomenon that has become more popular recently in mental health media known as *rejection sensitivity*. Sometimes, I perceive rejection even when I don't have definitive evidence for it. Fear of rejection is at the heart of the human experience. Figure 4.1 shows a picture my daughter drew depicting the emotional turmoil of rejection.

Ostracism threatens needs for belonging and leads to behavioral responses including attempting to recoup one's lost place in a social group, trying to regain control, or lashing out aggressively. Interestingly, ostracism also renders people more vulnerable to adopting extreme ideologies (Williams et al., 2019). However, in turn, ostracized individuals who subsequently join activist groups are, in turn, more likely to be ostracized by those outside of the activist group, particularly when the activist group stands for prosocial causes (Hales & Williams, 2019). This is critical to a key point of this book, that environmental advocacy can both nourish the soul and at the same time leave it prone to further hurt. Mental health clinicians will do best by their clients, particularly rejection sensitive ones, by encouraging advocacy while at the same time arming clients with tools for bolstering resilience to sustain prosocial efforts.

Environmentalism can trigger fear of social exclusion in subtle ways. I worry, for instance, about looking strange for bringing a non-disposable water bottle to barbeques, re-using plates, and avoiding straws. But feeling like you belong should not come at the expense of your environmental values. If we can be more accepting and inclusive, we can achieve true liberation that the latest gadgets, house extensions, and luxury items have not been able to give us. This is the liberation of acting out of compassion (for our bodies, the earth, and others) without fear of judgment and exclusion.

A related question is what potential risk might lie in being ostracized by those within the group with which you identify? Recall that despite the strengths of a collectivistic orientation, there is the potential risk of being discouraged from deviating from conformity. This means it also behooves activists to find reasons to join with others rather than to disagree over every lifestyle decision. Motivating people in holistic, value driven ways is more effective than making blanket demands. For example, someone who can afford a fancy car can be influenced to drive electric, while supporting the less well-off to advocate for federal funding for electric school buses.

There are many coping skills that have been proven to help quell the pain of exclusion, including simple measures such as holding a warm drink, as social exclusion is associated with lower temperature in fingers and warming one's fingers is associated with alleviation of the emotional pain associated with exclusion (Ijzerman et al., 2012) and viewing photos of animals like cats and dogs (Brown et al., 2016). These simple remedies are less costly to people's wallets as well as to the environment than more drastic means people take for inclusion such as plastic surgery.

By taking the initiative to include others, we help ourselves mend the pain of past exclusion. We prove to ourselves that we can join with others without rendering ourselves powerless through victimhood. We can live by values that help the earth, and this in turn can become contagious.

Next, we look at how FOMO contributes to the climate crisis but also challenges effective collaboration.

FEAR OF MISSING OUT (FOMO) AND CLIMATE CHANGE

As with other vignettes from my clinical work, this is an amalgamation of several clients' stories to protect confidentiality:

> During a recent session, Tara, a college graduate who had made the Dean's list every semester and got a well-paying job straight out of school, commented on how, despite being happy with her group of friends, she felt disconnected from them emotionally. Unbeknownst to her, this was a pivotal moment in her mental health journey because up until this point, our work had been focused on helping her dull the blade of academic perfectionism. Tara often sacrificed sleep and opportunities to socialize to overprepare for exams and did not settle for any grades below an A.

Tara's emotional disconnect from her friends could be framed in a way that helped her understand the downside of her obsessive work schedule, by highlighting that overworking was leading her to miss texts in her friend's group chat in which plans for get-togethers were arranged. This led to her missing out on conversations and outings even though her friends had every intention of including her.

Unfortunately, Tara found herself mending one problem, perfectionism, but experiencing a new one, the well-known phenomenon called The fear of missing out. Just like perfectionism, FOMO can lead to emotional stress, exhaustion and burnout when it becomes your primary motivation for, well, pretty much everything.

Social media is a FOMO exploiter. Whereas you might not have even been aware that your friend of a friend hosted a get together over the weekend, the photos on their platforms shove it in your face. Smart phone trackers lead to the same unsettling outcomes. All of this can exacerbate rejection sensitivity.

Here, we will explore two dimensions of FOMO and climate change. The first is the role that FOMO plays in exacerbating the ecological crisis by promoting consumerism and capitalism. The second focuses on how

FOMO influences environmental advocacy, motivating people to get involved but also preventing effective collaboration.

FOMO and Consumerism

In his seminal work *Constructing the Self, Constructing America*, psychologist Philip Cushman (1996) ties the idea of the self to psychology, history, and culture. He focuses on the fallout from World War II, which spurred an interest in manufacturing by gearing up to produce more ammunition. The post WWII era was a time when Americans were recovering emotionally from the traumas of the war.

The desire to fill a spiritual void was met with a growing interest in consumerism that was sparked by wartime production (Cushman, 1996). The intertwining of consumerism and identity was further entrenched by the nuclear arms race of the Cold War, which extended beyond weapons to a competition of which country (the United States versus Soviet Russia) had the most advanced technology and best gadgets. Cushman refers to this facet of the cold war as a "battle that focused on the wallet of the consumer" (p. 222).

The legacy of consumerism over all else is also echoed in the fact that Europe banned substances found to be toxic (such as lead and asbestos) well before they were banned in American products, the reason being the tremendous power of private companies lobbying the government to preserve their right to exploit Americans' health and well-being (e.g. Tabuchi, 2025). In the 1960s, with the help of widespread television viewing, consumerism led to the widespread purchasing of items for the sake of showing off wealth and technological advancement, with the possession of items (clothing, jewelry, etc.) becoming a way of expressing political views, personal freedom, spiritual transformation, and uniqueness (Cushman, 1996). In this way, psychotherapy has been unintentionally complicit in over-focusing on self-improvement. Obsession with celebrities is another manifestation of our attempt to fill our empty sense of self.

Cushman's (1996) work, which was published about 30 years ago, eerily holds even more truth in the 21st century, where people's push to reach celebrity status through social media and to water down life's authenticity by showcasing every "episode" on a personal YouTube channel or TikTok account has commodified every aspect of daily life. The message being inculcated from childhood is that for life and self to be meaningful, it must be recognized by others. The perceived need to commodify life entrenches us further in materialism, for now not only must we possess the latest gadget, but to be of any worth as a person, we must also possess others' attention and affection.

Here we come back to the theme of this chapter: How climate change relates to and interacts with our internal beliefs about ourselves and the world. If self-worth is tied to consumerism and competition, it means we are working late into the night (which means using more electricity) and buying the latest product, even if it is shipped by fuel-guzzling airplanes and trucks. When you are faced with the fear of not having worth, it is hard to embrace an alternative.

One of the toughest beliefs to crack in the therapy room is the idea of contingent self-worth, or that you are only worthy if you [have, produce] enough [success, wealth, attention, etc.]. When Carl Rogers coined the term *conditional self-regard* (Nevid, 2022) his intention was to inform parenting. For instance, in their efforts to discipline their children, parents should emphasize that they are correcting behavior, without criticizing the child as a person. While this advice still proves helpful to modern day parents, the reality is that they are up against a tsunami of toxic cultural messages, of which they have limited control and have become a part of themselves.

While Cushman's work (1996) indirectly speaks to FOMO, research has also investigated the relationship between FOMO and sustainability. A 2019 study (Kang et al., 2019) showed how aspects of FOMO, including *prestige sensitivity,* or a concern with one's image in others' minds, desires for connectedness and recognition, and concern about being left out lead to an uptick in consumption of brand items that are perceived as aiding in conforming to the norm.

FOMO and The Commodification of Climate Advocacy

How do I define myself as a climate advocate? This question has been on my mind since my daughter's birth triggered existential dread. The birth of both of my children within a three-year time frame coincided with emotional emptiness I started feeling in relation to my career. While I continued to find meaning in working with clients, I found myself feeling unfulfilled by the long hours and lack of community benefit my work offered at the time. Compared to Cushman's (1996) description of filling emptiness with addiction, whether to things, illicit substances or fame, turning to environmentalism instead seems ideal.

However, we cannot overlook a major obstacle in proactively responding to emptiness with activism. We need to remember that in coming together to solve the climate crisis, each of us is emerging from the competitive, self-absorbed, increasingly perfectionistic, isolation-fueling society that created this mess to begin with. We are trying to clean it up without knowing how to pull out the weeds. This comes across when people don't have time to volunteer for a climate agency or to compost, neither of which is likely to earn them recognition. When they do get invested, conflicts such as spitting matches over who gets to speak over a list serv can emerge.

While you don't need popularity to make a difference, self-efficacy is an important ingredient for success.

SELF-EFFICACY AND CLIMATE CHANGE

Social scientists describe *self-efficacy* as a belief in your own capability to reach a goal, including your ability to surmount challenges along the way. Self-efficacy is often used interchangeably with self-confidence, but some people have distinguished self-efficacy as being more task-specific (Artino, 2012), such as one's belief that they can accomplish a specific goal like earning a college degree. The term was originally coined by psychologist Albert Bandura as part of social cognitive theory, which includes the idea that we learn how to think and behave from what we observe in others' behaviors toward us and with each other. Bandura is well known for his Bobo doll experiments where young children were shown to model aggressive behavior after a contrived observation of adults beating up a Bobo (blow-up) doll (Nevid, 2022).

Similarly self-efficacy is a construct that can be heavily affected by our perception and interpretation of others. An example is the concept of Relation Inferred Self Efficacy (RISE) (Hsu et al., 2022). Research has shown that RISE can contribute to positive outcomes like greater effort, satisfaction, and helping others in the context of specific task-oriented goals.

A historical example of the power of believing in someone was Louis Leakey hiring Jane Goodall (Surujnarain, 2019) to study chimpanzees even though she didn't have a college degree at the time. Leakey wanted someone with an unbiased perspective, and he saw that Goodall was intelligent, observant, and passionate about understanding wildlife. Ultimately, Leakey encouraged Goodall to pursue her PhD in ethology.

Self-efficacy can impact how one responds to the climate crisis (Innocenti et al., 2023). For one, studying self-efficacy can help us predict whether anxiety will be manifested in adaptive, pro-active ways versus less adaptive ways. For example, one study (Qin et al., 2024) found that among middle and high school students in China, greater climate change anxiety was associated with fewer pro-environmental behaviors, contradicting research showing that adult climate activists tended to have a more anxious temperament. Interestingly, this relationship was mediated by *future self-continuity*, a concept that refers to seeing a stronger connection between your present and future selves. Specifically, the higher the anxiety of climate change, the more individuals paid attention to the present (and therefore less attention to their future selves), and therefore were less motivated to engage in legacy-creating pro-environmental behaviors.

In contrast, higher future self-continuity (Qin et al., 2024) was related to less climate anxiety and more pro-environmental behaviors. Among participants who were middle-school-aged, this effect was moderated by green self-efficacy, or one's confidence in planning and carrying out environmental goals and projects. The results of the study showed that the relationship between future self-continuity and pro-environmental behaviors existed under conditions of high green efficacy. The study authors attributed this finding to a need for identity inherent in this age group.

While research shows an important role for the individual's belief in themselves, self-efficacy does not grow in a vacuum. Our sense of how well we can overcome challenges is related to how the people we interact with are coping with the same obstacles. As Britt Wray (2022) describes in *Generation Dread: Finding Purpose in an Age of Climate Crisis*, our past failures can chip away at self-efficacy and make us vulnerable to learned helplessness, which goes hand in hand with beliefs that we will never have a meaningful effect on addressing a crisis as colossal as climate change.

Wray (2022) speaks to the importance of social context in describing how climate scientists have been disempowered from speaking to the full extent of the crisis by their perceived need to be seen as unemotional lest they come off as biased and therefore unscientific. Wray points out that the cold objectivity to which many climate scientists feel constrained inadvertently invalidates the strength of the public's outcry against the destructiveness of the fossil fuel industry.

Since Wray's (2022) book was published, an article in *The Guardian* (Carrington, 2024) revealed the despair felt by climate scientists, as they have voiced urgent calls for action that have been ignored by the government. This indicates reason to hope that the culture in academia is changing in a way that may be more relatable to the public. It also speaks to the concept of collective self-efficacy, which can be fostered in advocacy and activism groups that make efforts to celebrate successes however small.

In their book *The Psychology of Climate Change*, Geoffrey Beattie and Laura McGuire (2019) discuss self-efficacy in the context of explaining why public campaigns to increase awareness of the climate crisis have failed. Such advertisements aim to heighten fear of guilt in their audience without offering tools with which to combat the threat. Beattie and McGuire reason that the combined aim of heightening fear and reducing one's sense of personal control over the threat are likely to lower self-efficacy and to lead the audience to entrench themselves in denial and react with criticism, such as by believing that the messenger is being condescending.

Improving self-efficacy means moving away from an us-versus them mentality. It is what Peter Kalmus did for me when he assured me in his email response that I, as a non-scientist, am just as if not more powerful than the scientific expert in fighting for a livable earth (Communication with Peter Kalmus, March 31, 2022). This speaks to the concept of *participative efficacy* or the belief that one's individual contribution is important to the success of the entire group in achieving collective goals (Hamann et al., 2025).

Building self-efficacy might mean finding a way to praise your friend for making an effort to recycle even though you know recycling is an imperfect solution. You could then gently prompt them to maximize their impact by using fewer disposable resources to begin with.

In cognitive-behavioral therapy, self-efficacy is a facet of one's belief in themselves. Take, for instance, a person we will call Jerry, who grew up with highly critical parents who wouldn't allow him to do anything on his own. Jerry was born with an anxious temperament and, throughout his life, has been hypersensitive to threats in his environment. For example, if he sees someone walking toward him, he assumes the person is going to bother him.

A cognitive-behavioral clinician might take this, along with other information, to surmise that Jerry has beliefs that the world is a hostile place, that others are bound to dislike or criticize him, and that he is bound to attract criticism by inevitably failing any endeavor he pursues. Increasing self-efficacy can be accomplished in cases such as these by guiding the client to engage in pro-social behaviors, addressing negative biases in how they perceive the feedback they receive from others, and finding activities that bring them joy and a sense of accomplishment. In this way, the clinician and client work together to begin to chip away at old, harmful beliefs and pave the way for healthier ones.

An actual cognitive behavioral therapy (CBT) case conceptualization would be more elaborate, outlining mechanisms for problems being treated, such as avoidance or perfectionism, origins of these mechanisms from an extensive history obtained at intake, and precipitants of current problems, with the problems being further described as symptoms coalescing into a diagnosable disorder (Persons, 2008).

Just this snippet of Jerry, a conglomeration of many client stories, though, offers a window into how environmentalism can become complicated in someone with his history. A friend telling him that his efforts to recycle are for naught because "recycling is a sham" can trigger an underlying belief, learned through his childhood experiences and perceived through the lens of his threat-sensitive mind, that "nothing I do is good enough." This, in turn, can contribute to helplessness and resignation from pro-environmental behaviors.

We next turn to *justice sensitivity*, a specific kind of threat sensitivity that is pertinent to the climate crisis.

JUSTICE SENSITIVITY AND THE CLIMATE CRISIS

In a *Psychology Today* piece (Keluskar, 2024), I discuss the pros and cons of justice sensitivity, a concept that involves four components (Baumert & Schmitt, 2016): (1) being sensitive to perceived injustice; (2) reacting emotionally to it; (3) being preoccupied with it; and (4) feeling an urge to undo or fix it. Some forms of justice sensitivity are prosocial (Groskurth et al., 2023). They entail picking up on injustice being committed against others and are related to behaviors like altruism and activism (Baumert et al., 2022).

On the other hand, justice sensitivity can manifest as victim sensitivity, or believing you, personally, are being exploited (Baumert & Schmitt, 2016). Victim sensitivity can lead to aggressive behaviors, like vengeance seeking. It can also subtly hurt relationships. The three tips I offer for using justice sensitivity in good rather than harmful ways (Keluskar, 2024) can be applied to climate change psychology. First, I will list them below, and then we will dive into their relevance to eco-distress.

Justice Sensitivity Recommendation #1: Respond to Pain with Curiosity

Eco-Distress application: One aspect of the mental health toll of climate change is the pervasive invalidation implicated by the government's lack of action. Dialectical behavior therapist and licensed professional counselor Dr. Jesse Homan speaks to this issue in his application of Dialectical Behavioral Therapy to eco-distress (J. Homan, personal communication, November 4, 2023; Doherty et al., 2024). People who are alarmed about climate change, which is reasonable given the extremeness of the threat, are met with criticism for being over-reactive and emotionally dysregulated and are therefore invalidated. Less obvious, often unintentional acts of invalidation might come in the form of trying to soothe the person's angst.

Being faced repetitively with emotional invalidation makes people more sensitive to the threat of further invalidation. I realized that my own sensitivity to being invalidated contributed to feelings of insecurity when I interacted with activists. I assumed they were smarter, more assertive, and more organized. As a result, I chose to disengage from organizations because I sensed a threat that I was being seen as incapable even if it wasn't an actual threat.

The exercise of writing this book has in and of itself been a type of therapy for such environmental imposter syndrome. Following my own advice, I work on responding to my anxiety with curiosity as to what is behind my self-doubt when plenty of people in my personal and professional circles have encouraged me to pursue this project.

When it comes to preventing emotional insecurity from damaging collaboration with other environmental advocates, the work has to be on both parties' ends. We can start by recognizing that everyone concerned about climate change has dealt with collective invalidation. We can then move toward forgiving and lessening our judgment of allies.

Justice Sensitivity Recommendation #2: Keep a Kindness File

I have a word document where I keep a record of times I felt touched by others' kindness. I look back at it whenever I notice thoughts that reflect beliefs of others being hurtful or the world being unsafe or unfair.

Eco-distress application: This is something that could supplement a "media diet." While it is important to be aware of the formidable obstacles we face for climate action, there are many great efforts and innovations aimed at solving the problem, and it can help us sustain hope and prevent burnout by taking time to celebrate these wins. The book *Ongo: Everyday Nonviolence* (Cadden & Wiens, 2017), for instance, provides an exercise for creating mourning and celebration cards. One side of each card describes a situation you are mourning (e.g. the dismantling of the Environmental Protection Agency, EPA) and the other a win to celebrate (e.g. an initiative my town advertised a few years ago to collect used plastics and create park benches with them).

Justice Sensitivity Recommendation #3: Embrace the Diversity of Human Perception

Reminding ourselves of the many different ways people can perceive the same situation helps nurture compassion for ourselves and others.

Eco-distress application: Accepting and even embracing different points of view only partially applies here because in the case of climate change, we *are* being exploited by the greed of extreme capitalism and the lies and manipulative tactics of the fossil fuel industry. In this way, justice sensitivity can be adaptive, especially when we are focused on helping marginalized people that are even more exploited (think sweat shop workers, or the plight of residents of Flint Michigan during the city's water crisis).

However, embracing the diversity of human perception could also involve having empathy for people who haven't yet concerned themselves with the climate crisis. Perhaps their fear has entrenched them in denial. Perhaps there was a time when you yourself were either unaware of the risks of continuing to use and abuse natural resources unabated or you minimized the long-term consequences of these actions. Seeing things from others' perspective is the first step in taking steps to help more people understand what is at stake.

Where does that understanding lead us? While the quick answer is to support environmentalism, the specific paths to getting there are not always clear cut. While education about the environmental reality of lifestyle decisions can help us make better ones, there is always more to learn and something to acknowledge we don't know.

One of the reasons why even the most knowledgeable people experience blind spots is our tendency to process information through heuristics, or mental rules through which we filter situations to arrive at solutions that have worked in the past. Pertinent to climate change and environmentalism, research suggests that a cognitive bias known as the balancing heuristic (Sörqvist & Langeborg, 2019) leads people to believe they are compensating for the environmental damage they cause by purchasing goods labeled as "eco-friendly" without having enough information or knowledge to confirm they were successful.

In the short term, this might serve as a coping mechanism for the overwhelming extent of cultural change that will likely be necessary to sufficiently rewire society and culture. At the same time, we need to consider the long-term ramifications of the short cuts we take in how we think about the world. This leads us to our next discussion about how having an optimistic or pessimistic outlook affects our response to the climate crisis.

OPTIMISM, PESSIMISM, AND HOPE IN THE CLIMATE CRISIS

In *The Psychology of Climate Change*, Beattie and McGuire (2019) detail their own research on the subject of optimism bias, or the misguided idea that you will be among the lucky ones in even the worst of situations. Just like some people come to readily perceive threat, optimists seem primed to attend to stimuli that confirm their positive beliefs about themselves, the world, and others. While such an orientation can be protective against hopelessness, extreme optimism runs the risk of fueling complacency.

To investigate the effect of optimism bias on climate change beliefs, the researchers (Beattie et al., 2017, as cited by Beattie & McGuire, 2019) conducted eye tracking research and gave participants a questionnaire assessing their general, or dispositional, optimism. They found that optimists spent less time fixated on the content of climate change messaging regardless of the stance the content took on the issue. They spent even less time fixating on the "for climate change" article (i.e. that climate change is real, driven by human behavior, and will cause major problems that threaten human survival).

In contrast, non-optimists had longer fixations on the "for" climate change argument and framed their recall of the article with a focus on the root of human activity (Beattie et al., 2017, as cited by Beattie & McGuire, 2019). However, optimists framed the content of the article as a debate between two opposing points of view.

To further elucidate optimism bias as a cognitive phenomenon, the researchers (Beattie et al., 2017, as cited by Beattie & McGuire, 2019) grouped participants into three groups based on their level of optimism (optimist, medium-level, and non-optimist) and then asked participants the following as a measure of optimism bias (Beattie & McGuire, 2019, p. 47):

1. What is the probability of you personally being affected by climate change?
2. What proportion of people (living today) will be affected by climate change?
3. What proportion of future generations will be affected by climate change?

Participants who were higher in optimism rated themselves as less likely to be personally affected by climate change (Beattie et al., 2017, as cited by Beattie & McGuire, 2019), while other people were more likely and future generations were even more likely to be affected. In contrast, non-optimists had higher percentages for themselves being affected as well as higher percentages for others and future generations (although, like the optimists they rated themselves as relatively less likely to be affected than others and future generations). The authors concluded that realism is ideal in processing climate change messaging.

Realism is at the heart of a cognitive behavioral therapist's work. It is far easier and truthful to encourage a client to shift from thinking catastrophically to realistically than it is to get them to take a leap into extreme optimism.

At the same time, optimists are known for resilience. They take credit for the good things that happen to them and attribute negative events to the situation as opposed to over-personalizing them (Beattie et al., 2019). This increased resilience is much needed in the face of existential threat.

Where you fall on the spectrum of optimism is an ingrained part of your world view. It is not a switch you can simply turn on and off. Rather, what we need to focus on is how people who fall on different ends of the spectrum can collaborate effectively, balancing each other's viewpoint. Pessimists can give optimists a reality check and optimists can give pessimists a morality boost. But they have to be willing to listen to each other.

As a pessimist, I can find optimists to be annoying. I have wondered at how some of the most intelligent optimists I've known could so blatantly ignore certain obstacles and risks. This is a sign that I need to self-reflect on my own biases and ask myself a question I often ask my clients when they are faced with uncertainty: "What do you perceive would be the cost of being a little more optimistic about this situation?" The answer often reveals vulnerabilities such as fear of looking naïve or dumb, and this could lead to more conversations addressing these insecurities.

The key here is for each side to not get put off by the other, but to understand that each is processing the most basic components of their environmental stimuli differently and that the cumulative effects of such can render quite different worldviews, neither of which is one hundred percent "right" or "wrong."

Mutual understanding of these different worldviews can be achieved through a practice described in Radically Open Dialectical Behavior Therapy (RO DBT) called self-enquiry (Lynch, 2018) or finding questions that get you closer to an uncomfortable truth about yourself.

For example, as a pessimist, I could ask myself, *Is there a hidden sense of entitlement to my pessimism? In other words, do I assume I am all the wiser for knowing "the truth" about the world's darkness?* In contrast, the most optimistic relatives in my family also happen to be the most intelligent. I also ask myself if I imagine some hidden cost to optimism. Will I be called a fool if I am optimistic? Will being optimistic make me work less hard? This one brings to mind several clients who had the underlying belief, "If I am not stressed out, then I must not be working hard enough." These stories serve as reminders that I need to question the truth of my own assumptions about optimism and pessimism.

The wiser optimists in my life have come to realize that their rays of sunshine can unintentionally come off as invalidating to another's shadows. They can also serve as a veil of denial, blocking painful but important conversations. Such wiser optimists don't stop being optimists. Rather they build a tolerance for containing others' negativity and gently offering their perspective while validating an understanding of the darker perspective.

Such an offering can be translated effectively into constructs that can be taught more readily than optimism itself, such as hope and faith. As detailed by environmentalists like Joanna Macy and discussed by Jane Goodall, the concept of active hope implies an intentional contribution, however small, toward a solution to the climate crisis (Park et al., 2020). It is the idea of adopting a positive motivational state and sense of agency to pursue a goal however uncertain the outcome may be.

In cognitive behavioral therapy, I talk to my clients about the concept of active hope to help them cope with uncertainty. We start off with a discussion about acceptance skills from dialectical behavior therapy (DBT; Linehan, 1993), which refer to seeing reality for what it is without trying to deny, fix, or fight against it (more on DBT in Chapter 7). It is acceptance without necessarily liking or loving current reality and it does not negate the possibility of future change.

I ask clients whether, given that the situation is the way it is at present, they can carve at least a tiny sliver of hope toward feeling better, more empowered, more purposeful, etc. in the future. At times, what starts off as a conversation about coping with the uncertainty of a job or financial prospects evolves into disclosure of the client's eco-distress. Whereas blind faith risks silencing open, vulnerable discussions of climate fears, active hope normalizes them, paving the way for validation and collective action.

Hope is one target of the positive psychology movement, which has been the zeitgeist of mental health for about 25 years, but has been criticized more recently. As opposed to more traditional views of psychology, which focus on pathology and understanding people's misery, positive psychology (Steger, 2024) focuses on how to bring about or enhance positive psychological experiences such as happiness, resilience, gratitude, kindness altruism/generosity, and longevity. We next explore positive psychology and how it feeds into another area called regenerative positive psychology.

Regenerative Positive Psychology

Widespread access to media sources such as the internet and social media has boosted awareness of unprecedented crises such as climate change, political divisiveness, and inappropriate use of artificial intelligence (AI). These tensions are addressed in a 2024 paper about *Regenerative Positive Psychology* (Steger, 2024), according to which psychology has over-focused on individual well-being when it needs to broaden its scope to repairing global problems including ecological crises. The field of clinical psychology could shift its focus the effect of global problems on people because our capacity to improve individual well-being will be limited in a broken world.

In his thesis on *Regenerative Positive Psychology*, Steger (2024) writes:

> It is as if we are earnestly working beside each other in a sprawling garden, tending carefully to the small plots within reach, while wildfires, tornados, floods and enraged mobs decimate acre after acre, hectare after hectare at the garden's edges, ever approaching our isolated, fragile plants. (p. 386)

These words offer a metaphor for eco-distress in portraying the sense of impending doom and the frustration that comes from observing those who are so absorbed in their little gardens that they seem unaware of the encroaching catastrophes. These others, on the other hand, might see us as oblivious for seemingly ignoring their pretty, peaceful gardens.

Steger (2024) points out that while promoting a relationship and connectedness with nature is beneficial to both psychological well-being and has been linked to more pro-environmental behaviors, it is also important to have a reciprocal relationship with nature in which we care for it in addition to it sustaining us.

One core principle of regenerative psychology (Steger, 2024) is that happiness is best achieved through moderation. Moderation as a value and behavioral principle stands to challenge the unhealthy underlying belief of "never good enough," a belief that extreme capitalism reinforces and perfectionism feeds, whether it applies to physical appearance, which leads to the consumption of products to help us feel more attractive, or performance, leading to relentless drives to outperform others at all costs. A spiritual orientation toward moderation (to be explored further in Chapter 5) can help reframe the fear of deprivation resulting from pro-environmental behaviors, to one of contentment.

The idea of a balanced existence is intuitively appealing and yet for many feels unachievable due to the realities of our social order. How, for instance, does a single parent balance work, family demands, and rest when she is expected to do everything herself? A life that is not balanced leads to behaviors that are also out of balance. It is the reason I continue to have items shipped to me. Even with a helpful partner, the demands of my life make the more convenient and often less environmentally friendly options irresistible. The rat race that we have established, though well intentioned to promote equality and opportunity, has not been successful at doing so.

An example from ninth grade comes to mind here. At lunch, I sat with a group of girls who were in advanced classes. Every time we had an exam returned, they would take turns announcing their grade. When I told them I got an 89 on a math test, this was met with a fake frown, and a "Sorry to hear" in a condescending tone. I was so upset by the grade and my peer's reaction that I told my therapist at the time, who responded by arguing with me that my *needing* to get in the 90s was "irrational." Well, of course, it is, I thought, but *you* try sitting with the comparison clique every day. You try being OK with what most people outside of my clique would see as a good enough grade when the only social circle I was included in saw it as most definitely not OK.

Years later, when I was on the other side of the therapy room counseling adolescent overachievers, I tried to take a somewhat different approach, encouraging them to be their own activists by taking the rebellious stance of being OK with imperfection.

Theoretically, regenerative positive psychology would target multiple overlapping health systems, so that the same messages are getting across to the community through schools, medical offices, and religious communities. We could engage stakeholders, such as having older teens counsel younger children about the limitations and dangers of perfectionism, and the joy that an in-person hang out with friends can bring, as opposed to scrolling each other's social media feeds. All of this could lead to a more just, kind, and environmentally friendly world.

CONCLUSION: IS ECO ANXIETY ADAPTIVE?

Eco-distress is a valid response to an ecological crisis, not a pathology. I stress that the point of this book is not to use psychotherapy as a pacifier for climate change distress. Although eco-distress motivates climate solutions, like all other types of anxiety, it comes with baggage that can keep us from living in accordance with our values. Rumination over whether a fellow climate advocate judged me for using a plastic cup might lead me to avoid the next environmental group meeting she is bound to attend.

One of the reasons CBT has proven so efficacious is its unique ability to directly target underlying, maladaptive beliefs people hold about themselves, their relation to others and the world, and their future. By acknowledging the role that society and consumerist culture play in maintaining these beliefs, and by treating such societal ills as part of treating the whole person, we can direct people toward both a sense of purpose and a place in helping to prevent the worst consequences of climate change.

REFERENCES

Artino, A. R. Jr. (2012). Academic self-efficacy: From educational theory to instructional practice. *Perspectives on Medical Education, 1*(2), 76–85. https://doi.org/10.1007/s40037-012-0012-5

Baumert, A., Adra, A., & Li, M. (2022). Justice sensitivity in intergroup contexts: A theoretical framework. *Social Justice Research, 35*, 7–32. https://doi.org/10.1007/s11211-021-00378-9

Baumert, A., & Schmitt, M. (2016). Justice sensitivity. In C. Sabbagh, & M. Schmitt (Eds.), *Handbook of social justice theory and research* (pp. 161–180). Springer. https://doi.org/10.1007/978-1-4939-3216-0_9

Beattie, G., Marselle, M., McGuire, L., & Litchfield, D. (2017). Staying over-optimistic about the future: Uncovering attentional biases to climate change messages. *Seiotica, 218*, 21–64.

Beattie, G., & McGuire, L. (2019). *The psychology of climate change.* Routledge.

Beck, A. T., Rush, J. A., Shaw, B. F., & Emery, G. (1979). *Cognitive therapy of depression.* Guilford Press.

Brown, C., Hengy, S. M., & McConnell, A. R. (2016). Thinking about cats or dogs provides relief from social rejection. *Antrhozoös, 29*(1), 47–58. https://doi.org/10.1080/20414005.2015.1067958

Cadden, C., & Wiens, J. (2017). *The Ongo book: Everyday nonviolence.* Baba Tree.

Carrington, D. (2024, May 8). We asked 380 top climate scientists what they felt about the future... *The Guardian.* https://www.theguardian.com/environment/ng-interactive/2024/may/08/hopeless-and-broken-why-the-worlds-top-climate-scientists-are-in-despair

Chwialkowska, A., Bhatti, W. A., & Glowik, M. (2020). The influence of cultural values on pro-environmental behavior. *Journal of Cleaner Production, 268,* 122305. https://doi.org/10.1016/j.jclepro.2020.122305

Clifton, J. D. W., & Meindl, P. (2021). Parents think – incorrectly – that teaching their children that the world is a bad place is likely best for them. *The Journal of Positive Psychology, 17*(2). https://doi.org/10.1080/17439760.2021.2016907

Cushman, P. (1996). *Constructing the self, constructing America.* Grand Central Publishing.

Doherty, T., Artman, S., Homan, J., Keluskar, J., & White, K. (2024). Environmental identity-based therapies for climate distress: Applying cognitive behavioural approaches. *The Cognitive Behaviour Therapist, 17,* e29. https://doi.org/10.1017/S1754470X24000278

Eom, K., Kim, H. S., Sherman, D. K., & Ishii, K. (2016). Cultural variability in the link between environmental concern and support for environmental action. *Psychological Science,* 1–9. https://doi.org/10.1177/0956797616660078

Fallahnejad, A. (2025). The philosophy of wabi-sabi in Japanese culture and its effects on modern lifestyle for young Japanese [book review]. *Journal of Academic Ethics, 1,* 1–11.

Groskurth, K., Beierlein, C., Nießen, D., Baumert, A., Rammstedt, B., & Lechner, C. M. (2023). An English-language adaptation and validation of the justice sensitivity short scales–8 (JSS-8). *PLoS ONE, 18*(11), e0293748. https://doi.org/10.1371/journal.pone.0293748

Hales, A. H., & Williams, K. D. (2019). Extremism leads to ostracism. *Social Psychology, 51,* 3. https://doi.org/10.1027/1864-9335/a000406

Hamann, K., Junge, E., Blumenschein, P., Dasch, S., Werne, A., & Bleh, J. (2025). *The psychology of collective climate action: Building climate courage.* Routledge.

Hsu, W., Shang, I., Pan, Y., & Chou, C. (2022). Students' efficacy profiles and outcomes of perceived relation-inferred self-efficacy support in physical education. *International Journal of Sport and Exercise Psychology, 23*(1), 56–69. https://doi.org/10.1080/1612197X.2022.2043926

Ijzerman, H., Gallucci, M., Pouw, W. T. J. L., Weißgerger, S. C., Van Doesum, N. J., & Williams, K. (2012). Cold-blooded loneliness: Social exclusion leads to lower skin temperatures. *Acta Psychologica, 140*(3), 283–288. https://doi.org/10.1016/j.actpsy.2012.05.002

Innocenti, M., Santarelli, G., Lombardi, G. S., Ciabini, L., Zjalic, D., Di Russo, M., & Cadeddu, C. (2023). How can climate change anxiety induce both pro-environmental behaviours and eco-paralysis? The mediating role of general self-efficacy. *International Journal of Environmental Research and Public Health, 20*(4), 3085. https://doi.org/10.3390/ijerph20043085

Kang, I., Cui, H., & Son, J. (2019). Conforming consumption behavior and FoMO. *Sustainability, 11*(17), 4734. https://doi.org/10.3390/su11174734

Keluskar, J. (2023, July 7). The psychology of climate change. *Alone Together.* https://www.psychologytoday.com/us/blog/alone-together/202307/the-psychology-of-climate-change

Keluskar, J. (2024, September/October). Are you sensitive to injustice? *Psychology Today, 57*(5), 22–23.

Linehan, M. (1993). *Cognitive-behavioral treatment of borderline personality disorder.* Guilford.

Litz, B. (2022, September 22). *Moral injury: An overview of conceptual, definitional, and treatment issues* [Webinar]. Association for Behavioral and Cognitive Therapies.

Lynch, T. (2018). *Radically open dialectical behavior therapy: Theory and practice for treating disorders of overcontrol.* Context Press.

Naiman, S. M., Stedman, R. C., & Schuldt, J. P. (2023). Latine culture and the environment: How familism and collectivism predict environmental attitudes and behavioral intentions among U.S. Latines. *Journal of Environmental Psychology, 85*, 101902. https://doi.org/10.1016/j.jenvp.2022.101902

Nevid, J. S. (2022). *Essentials of psychology, concepts and applications* (6th ed.). Cengage.

Park, A., Williams, E., & Zurba, M. (2020). Understanding hope and what it means for the future of conservation. *Biological Conservation, 244*(108507), 1–11. https://doi.org/10.1016/j.biocon.2020.108507

Persons, J. B. (2008). *The case formulation approach to cognitive-behavior therapy*. Guilford Press.

Qin, Z., Wu, Q., & Bi, C. (2024). The relationship between climate change anxiety and pro-environmental behavior in adolescents: The mediating role of future self-continuity and the moderating role of green self-efficacy. *BMC Psychology, 12*, 241. https://doi.org/10.1186/s40359-024-01746-1

Samarasinghe, R. (2012). The influence of cultural values and environmental attitudes on green consumer behavior. *International Journal of Behavioral Science, 7*(1), 83–98. https://doi.org/10.14456/ijbs.2012.6

Sörqvist, P., & Langeborg, L. (2019). Why people harm the environment although they try to treat it well: An evolutionary-cognitive perspective on climate compensation. *Frontiers in Psychology, 10*(348), 1–5. https://doi.org/10.3389/fpsyg.2019.00348

Sparkman, G., Geiger, N., & Weber, E. U. (2022). Americans experience a false social reality by underestimating popular climate policy support by nearly half. *Nature Communications, 13*, 4779.

Steger, M. F. (2024). Regenerative positive psychology: A call to reorient wellbeing science to meet the realities of our world. *Positive Psychology, 20*(3), 373–396. https://doi.org/10.1080/17439760.2024.2365259

Surujnarain, R. (2019, August 7). *Leakey and Goodall: Scientists who changed how we refine 'human.'* Jane Goodall's Good for All News. https://www.news.janegoodall.org/2019/08/07/7138/

Tabuchi, H. (2025, June 16). E.P.A. Plans to reconsider a ban on cancer-causing asbestos. *The New York Times*. https://www.nytimes.com/2025/06/16/us/epa-trump-asbestos-ban-delay.html

Williams, K. D., Hales, A. H., & Michels, C. (2019). Social ostracism as a factor motivating interest in extreme groups. In S. C. Rudert, R. Greifeneder, & K. D. Williams (Eds.), *Current directions in ostracism, social exclusion, and rejection research* (pp. 18–31). Routledge/Taylor & Francis. https://doi.org/10.4324/9781351255912-2

Wray, B. (2022). *Generation dread: Finding purpose in an age of climate crisis*. Knopf Canada.

Chapter **5**

Climate Change and Spirituality

THE INTERCONNECTEDNESS OF NATURE

The renowned fiction writer, Isabel Allende, speaks to an interconnectedness in nature that brings to mind how spirituality relates to the climate crisis. Twenty years after reading *And of Clay Are We Created* (1989), one of Allende's literary gems, I realize how much the story relates to our struggle with the vicissitudes of nature, only worsened by escalating global temperatures. The story is fictional but is based on a true event and a news story about a young girl who was trapped in mudslides resulting from a volcanic eruption in Colombia in 1985 (Stone, 2023).

And of Clay Are We Created (1989) is about a 13-year-old girl, Azucena (by no accident, the name translates to the name of a flower, Lily), who is trapped after a mudslide ravages her town. Her head and arms are above ground only because of a tire that is keeping her from getting sucked below.

Azucena is followed by journalist Rolf Carlé, who makes her an international sensation. Carlé makes valiant efforts to save her but to no avail, realizing that without the proper equipment, it will be impossible to free her. As they wait days for the pump that would free Azucena from the jaws of the clay earth, irony unfolds as the television equipment for Carlé's filming is ever present at the scene, an attestation to the privileges of higher income countries as they stand and watch the needless suffering of the economically disadvantaged. The story ends tragically with Azucena dying as she sickens while stuck in the mud. Carlé, though, doesn't get away with a pain-free experience. In addition to watching the young girl suffer and die, Azucena's situation triggers traumatic memories of his family living through the Holocaust and the Nazi occupation of Austria, where he grew up. Allende is able to unite the opposing universes the

DOI: 10.4324/9781003489665-6

characters represent through their shared human fragility in the face of forces beyond their control.

Allende's work conveys how the synergy of science and spirituality can hold the key to solving our planetary crisis.

SPIRITUALITY AS RESILIENCE

While Allende doesn't believe in ghosts or spirits, she started infusing her writing with spiritual accents in *The House of the Spirits*, her first well-known novel that was based on the death of her grandfather (Beard, 2016). Her connection with the spiritual realm was strengthened by the untimely death of her daughter Paola, a psychologist and educator who provided humanitarian support to impoverished communities. Allende has said that she strongly feels her daughter's presence, and this has motivated her to imbue her characters with the ability to gain wisdom from those who have passed (Cruz, 2025).

In interviews about her writing, Allende talks about a concept known as *organic justice* (Dreyfus, 2023) or *natural justice* (Hoffman, 2017), which comes from a sense of fairness and authenticity in the community, as opposed to institutional legal systems.

One of the many examples highlighting environmental devastation as a social justice issue is the exploitation of indigenous people's land for profit. Take, for example, President Biden's support of the Willow Project (Ahtuangaruak, 2022), a large-scaled oil extraction project in Alaska that will destroy habitats that local indigenous communities, such as Niusquit's Inupiat Eskimo people, rely on for vital resources such as the fish and caribou they hunt for food, as well as polluting their air and water.

While climate change is fundamentally related to science, we need to give voice to cultures whose practices are less science-based in order to gain their trust in collective environmental action. Additionally, we can stand to learn from their sustainable practices, as these represent a vital complement to the scientifically based solutions currently being proposed.

The idea of a collaborative relationship with indigenous people is described in a paper by Johannes Luetz and Patrick Nunn (2023), *Spirituality and Sustainable Development: An Entangled and Neglected Relationship*. Luetz and Nunn (2023) argue that many sustainability initiatives fail because they do not consider conflicting spiritual and cultural practices. This is a stark reminder that enforcing practices on others because they are "in their best interests" can be a show of veiled colonialism in conveying the false notion that sustainability must be done the Western way and no other.

Take for instance, the sustainable act of choosing to purchase an electric car over a gas vehicle. This has been one of my chosen venues for sustainable action, but also an expensive one that I have the privilege to be able to choose.

The authors (Luetz & Nunn, 2023) highlight a continuum of nature-based values ranging from those in which humans are seen as part of a vast, interconnected web of life (ecocentric) versus the standpoint that natural resources exist for human benefit and use (anthropocentric). People's values fall on a continuum between these two extremes. Someone from a strongly eco-centric orientation will be more likely to see the purchase of an electric vehicle as a mega-usurpation of natural resources, or even worse, an exorbitant investment in guilt protection.

My intention is not to shame myself and others whose sustainable actions run along more anthropocentric lines. Rather, I wish to underscore the importance of showing Radical Openness (an RO DBT concept that will be explored in a later chapter) to other world views, which involves opening oneself up to the idea of learning from others, even those whose views might seem fundamentally different or at odds with your own (Lynch, 2018).

In cultivating a reverent stance toward the natural world, indigenous cultures nurture an interconnected relationship with nature that enhances motivation to preserve it. Importantly, individuals from indigenous cultures are more apt to derive a sense of meaning and purpose through the role of steward versus conqueror of nature (Luetz & Nunn, 2023).

Research has shown that helping people foster a stronger connection with the natural world can increase prosocial behaviors such as cooperation (Dopko et al., 2019) and inspire a sense of belonging that one might be lacking in their human relationships. Spending time in nature has been shown to directly affect the nervous system (Nisbet et al., 2019) in a way that contributes to improving mood.

Researchers at Massachusetts Institute of Technology (MIT; Bell et al., 2025) found that when participants were exposed to images of nature, they showed changes in brain wave activity that corresponded with improved performances on cognitive tasks requiring focus. More research is needed to clarify whether the brain changes directly caused performance improvements, as these improvements could have been influenced by practice effects. Also, the results did not consistently show a difference in performance changes in participants who viewed photos of nature and those who viewed urban photos, nor were changes in mood found in response to either type of photo.

However, it is worth noting that positive mood changes were seen in response to the nature photos when participants had higher levels of a personality trait called nature relatedness, or the extent to which someone sees their relationship with nature as an important part of their identity (Bell et al., 2025).

This suggests that it is important to consider people's *nature connectedness*, a psychological construct that reflects people's attitudes and relationship with nature (Sheffield et al., 2022) to ensure that nature exposure is mood boosting.

Knowing that the nature-related thoughts, feelings, and behaviors one brings to the table influence how likely they are to find solace in nature, and in so doing, feel inspired to conserve it, begs the question of whether it is a quality that can be strengthened, as opposed to being unchangeable.

The existing research on this matter has revealed promising results. For example, a 2022 meta-analysis (Sheffield et al.) showed that interventions can be used to increase this quality. These interventions included providing indirect contact with nature, such as asking participants to look at photos or videos, as well as more direct contact such as visiting an animal park.

Nature connectedness and the sense of wonder it evokes contribute to nature being a source of spirituality that nourishes the soul. In contrast, the media sends the message that emotional wellness is tied to fighting against nature as opposed to embracing it, as we see through social media's spread of sensational age-reversal trends being used by people who are as young as their teens and 20s (Doyne, 2024).

Spirituality can be distinguished from organized religion. The role of religion in organizing human action is discussed next.

RELIGION

Religion evokes mixed emotions. For some, it is associated with the comfort of community, while for others it is a vehicle of discrimination brought about by rigid rules and beliefs.

In the context of addressing the climate crisis through the vehicle of religion, we should consider that the principles of some religions are more conducive to environmental action. For example, reducing waste is a core principle of Jainism (ARC, 2023). Modern witchcraft, or Wicca, emphasizes sustainability and respect for nature and the environment (Tejeda, 2025). Hinduism is rooted in beliefs of nature being sacred (Singh et al., 2021). While all religions share similar themes, we can also learn much from understanding their differences, particularly along the eastern-western divide.

As Stephanie Kaza (2008) details in her book *Mindfully Green*, the four noble truths of Buddhism can be applied to states of mind people might experience in response to climate change. Examples of such mind states include anger at government inaction, anxiety about the catastrophic consequences of environmental devastation, and uncertainty about what exactly is in store for us. While such states are adaptive in motivating action, Kaza argues that they distract people from the flexible way of thinking that is needed to adeptly address the climate crisis.

The four noble truths of Buddhism (Quyet et al., 2022) are as follows:

Dukkha: The idea that suffering is an unavoidable part of existence.
Samudaya: Suffering results from attachment to what is desired as well as actions arising from negative mental states.
Nirodha: We can end suffering by letting go of attachment to what we crave.
Marga: The truth of the eight-fold path that leads to ending suffering.

Kaza (2008) guides the reader in responding effectively to negative mental states via the wisdom of these truths. She shares that the first step is to be aware of negative states of mind and their triggers (or hooks). Then, you can use mindfulness to guide your mind back to the present moment, such as reminding yourself that no person as an individual carries the entire responsibility of solving climate change and global warming. Many other people and nations are taking action. The next step is to cultivate positive mind states through practices such as meditation or prayer.

Religions both Western and Eastern share common principles that could be applied to conservation. Most religions, for instance, practice temporary fasting for special events. The spirit of restraint can be applied to conserving earth's resources. All involve self-reflection and distancing oneself from their ego to consider the common good.

However, the humanitarian efforts for which most religious institutions are known bring about mixed responses. While there is great potential for religion to facilitate the social connections and collective action that are needed to address climate change, there is another side to them that brings to mind the words of American novelist and non-fiction writer Anne Lamott, "… Help is the sunny side of control" (Dreyfus, 2024, 23:35–23:38).

A real-life and extreme example of Lamott's words is portrayed in the disturbing HBO documentary *Savior Complex* (Jesko, 2023), which brings to light the complicated role of religion in social justice issues. The documentary tells the story of Renee Bach, a young American woman who volunteers to help the poor and hungry in Uganda as part of her church's missionary. She finds purpose helping others and ultimately is able to create a hospital. However, her efforts go too far when she administers intravenous medications to frail children despite having no medical training. This leads to accusations of her being responsible for the deaths of possibly over 100 children due to over-feeding and other misjudgments in treatment, including going against a medical doctor's orders, opting instead to go with a gut feeling in a case that ended in the child dying.

The miniseries (Jesko, 2023) goes on to discuss an organization known as *No White Saviors* that aims to prevent uncredentialed white volunteers from taking advantage of vulnerable African people, an act that, in reality, represents an extension of colonialism in seeking to control the "other" under the guise of helping. Ultimately, after *No White Saviors* investigates the case and empowers victims to take Bach to court, Bach is forced to leave the country and is condemned on social media.

A theme echoed throughout the series is Bach's belief that she was called by God to intervene and save children's lives. She even adopts one of the children she cares for, raising her in the United States.

The belief that God is calling you to take what is not yours is one danger posed by religion, also exemplified by Biblical interpretations justifying horrific acts such as slavery, colonization, and the subjugation of women.

At the same time, religion is a powerful vehicle for community, and organizations such as Earth Care Congregations take a stewardship role with nature (Presbyterians for Earth Care: An Eco-Justice Network, 2020). Earth Care Congregations (ECC) is a program that encourages Presbyterian churches to act in earth-friendly ways. In its *Guide to Greening Presbyterian Churches* (Version 5, 2025), ECC explains that congregations can receive various honors, including certification as an Earth Care Congregation, which confers a certificate, use of the ECC logo, listing on the ECC website, a press release for local publicity and other resources for pledging to care for God's earth in four categories: Worship, education, facilities, and outreach. ECC (Version 5, 2025) states that its mission is tied to eco-justice in defending all of creation and helping those who are the most vulnerable to environmental devastation.

People can also influence religious institutions to incorporate environmentalism into their projects. Lauren Krueger, a mother of two young children, spearheaded climate actions through her role as a board member of her local synagogue in Merrick, Long Island. She started a sustainability committee at Merrick Jewish Center and has facilitated pro-environmental shifts within the culture of her religious community, such as organizing a Sustainability Shabat. She collaborated with an organization called Dayenu (Dayenu: A Jewish call to climate action, 2026), which is a movement of American Jews who are committed to fighting the climate crisis.

Another example of the power of religious followers is the extension of the Parachutes for the Planet program to the Pope's visit to Panama City (Goldman, 2022). In January 2019, students from Catholic parish schools decorated parachutes with words and drawings about their feelings on climate change. These parachutes were displayed during the Pope's visit to Panama City. The Pope took one parachute back to the Vatican and soon after the trip, declared a climate emergency for the Catholic Faith. As stated by Linda Goldman (2022) in her book *Climate Change and Youth: Turning Grief and Anxiety into Activism*, "The parachutes could have been a catalyst to bring about this strong declaration" (p. 182).

On the other side of the coin, when it comes to constituents influencing the interests of religious organizations, there is also the potential for harm. A case in point is the Christian Evangelicalism of the American far right. As discussed in a 2022 *New York Times* piece entitled *The Growing Religious Fervor in the American Right: 'This Is a Jesus Movement'* (Dias & Graham, 2022):

> The Christian right has been intertwined with American conservatism for decades, culminating in the Trump era .... But worship ... was largely reserved for the church. Now, many believers are importing their worship of God, with all its intensity, emotion and ambitions, to their political life. (para. 5)

There is also potential for governments to cover up complacency with the compassionate tongue of religious sentiment. An example of this is caving into the gun-lobby's adoption of the saying "thoughts and prayers" in response to mass shootings rather than standing up to pro-gun lobbying and taking meaningful action to prevent mass violence (Hassan, 2019). In the words of Rev. Deanna Hollas, who became the first minister of gun violence prevention in the United States, "While all that we do as Christians should be rooted in worship and prayer, it should not stay there. It is like breathing – worship and prayer is the in-breath, and action is the out-breath" (para. 3). One could say that constitutional amendments have taken on a black and white dimension when the right to bear arms applies to everyday citizens being able to acquire military arsenal.

Similarly, religion, when taken to extremes can be molded to fit any ideology, harmful or helpful, and therefore has the potential to hurt individual rights, whether they be reproductive, environmental, or more broadly, equal protection under the law, as dictated by the 14th Amendment of the US Constitution.

It goes without saying that many turn to prayer and religious teachings for guidance and solace. Seeking psychological counseling from a religious leader is the norm for many people, whether for cultural reasons or due to lack of access to mental health professionals. However, for others, spirituality, distinct from a religious institution, figures more prominently in emotional healing and wellness.

Spirituality also plays a prominent role in our relationship to nature to the extent that an Ecospirituality Scale was developed and validated (Suganthi, 2019). The scale assesses five domains: "Dwelling," which entails reflecting on the things in the universe and our meaning and purpose as part of it; "Caring," which focuses on how we care for the environment and finding meaning and purpose in nurturing it; "Revering" emphasizes awe and gratitude for living in the universe and promoting sustainability; "Experiencing," which relates to taking pleasure in the beauty of the universe; and "Relating," which entails a sense of mystery in being part of the universe.

One study, using a different measure of ecospirituality, found that it predicted moral concern for nature, independent of religious or political identity, and yet influenced decisions involving a nature-economic trade-off as well as political voting (Billet et al., 2023). We next turn to spirituality and its effect on both mental health and responses to climate change.

SPIRITUALITY APPLIED IN CLINICAL PRACTICE

We began this chapter with a discussion of spirituality as it relates to our connectedness with all living things, from other humans to the trees in our yard. In addition to the dimension of inter-relatedness, spirituality can be defined by our willingness to submit to a power greater than ourselves, whether it be a divinity, nature, the at times seemingly random, chaotic energy of human whims, or even simply our intuition.

Unfortunately, spirituality can be channeled through harmful vehicles of expression. For example, research by Frenken and Imhoff (2025) found that conspiracy beliefs, authoritarianism, and spirituality tend to co-exist in people and that these attitudes independently predict willingness to support illegal and violent protests, but this is particularly so when authoritarianism is combined with spirituality.

How do we steer spirituality toward climate-friendly and humanitarian beliefs and actions? In this section, I make the argument that we can't without inspiring a spirituality that involves a basic trust in each other.

Spirituality is not everyone's jam. While some clients have found it to be an appealing alternative to religion, the institutions of which have historically been oppressive and even traumatic, others have trouble untying the idea of appealing to a higher power from whichever form that higher power took in the religion they were raised in. However, according to a Pew Research study (Van Kessel et al., 2018), 20% of Americans endorse spirituality and faith when describing what provides them with a sense of meaning.

Cognitive behavioral therapy (CBT) is a goal-oriented psychotherapy that focuses on building clients' coping skills toward the effort of helping them manage psychological symptoms like anxiety and depression. Thus, efforts to incorporate religion are mostly client-initiated. For example, if a client derives comfort from prayer, praying can be listed as a coping strategy during times of stress. If they derive a sense of belonging and companionship from their church community, encouraging them to attend church regularly can be incorporated into behavioral activation, an intervention for clients struggling with depression that encourages them to plan and schedule routine activities that improve mood (e.g. Stein et al., 2021).

In contrast, the broader term of spirituality is implied in CBT interventions such as dialectical behavior therapy (DBT) (Linehan, 1993), through meditation and mindfulness in nature-based practices, such as observing the various textures, sounds, and colors of the wild while walking through the forest. It is also implied in the concept of wise mind (a synthesis of intuitive emotional mind and logical rational mind) and in encouraging prayer to a higher being as a potential coping skill.

Spirituality also applies to stoicism (Irvine, 2019), which might come as a surprise given the stereotype of stoicism as cold and logical. However, many of my clients have benefited from the stoic challenge which is the idea that rather than seeing an obstacle as a threat, which is more likely to lead to "Why me?" thoughts and accompanying helplessness, it could be perceived as a challenge from a higher power, providing the opportunity to rise to the occasion. I have personally found this coping skill to be useful in parenting, as it goes hand in hand with the common conclusion that my children are "testing me," but helps me feel more efficacious within that dynamic.

In addition to direct CBT skills such as mindfulness and meditation, and indirect skills such as the stoic challenge, spirituality can be utilized to warm up a hesitant client to the idea of applying CBT strategies for making changes in their behavior. An example would be seeking insight into their behaviors, thoughts and emotions from symbols that they imagine are presented by a spiritual deity.

In *Mindfully Green*, Kaza (2008) notes that bringing spirituality into climate action includes doing whatever pro-environmental action excites you. In this sense, many elements of CBT are spiritual, such as creating feelings charts with children that reflect their interests, like characters from *Minions*, *Star Wars*, or dinosaurs displaying different emotions.

As I read Kaza's (2008) work, I drew connections between her principles and concepts from several CBT interventions, including DBT, RO DBT, and acceptance and commitment therapy (ACT). I list these correlates in Table 5.1 for the reader to consider and explain my reasoning below. However, I should note that these parallels are hypothetical and based on my observations alone. I am unaware of any research that has confirmed correlations among these constructs.

Seeking Wisdom Sources

Kaza (2008) notes that environmental caregiving is a lifelong task and that it is important to seek inspiration and encouragement from wisdom sources, or individuals who are further along what she refers to as the green practice path. In turn, you pass down your wisdom to others who are at earlier stages

Table 5.1 Cognitive-behavioral therapy correlates of concepts from Mindfully Green

Mindfully Green *Concept*	*CBT Correlate*
Seeking Wisdom Sources	DBT concept of Wise Mind; RO DBT concept of therapist as ambassador
Systematic Seekers	ACT concept of living in accordance with values
Serendipitous Seekers	RO DBT concepts of Radical Openness and Flexible Control
Shenpa	Urges to engage in compulsive behavior, like doomscrolling; compulsive behaviors represent an overcontrolled coping mechanism (a concept from RO DBT) that is a response to tension or emotional discomfort
Shenlok	Resisting urges to cave into pleasure-seeking impulses associated with the discomfort of Shenpa
Green Lifeway	RO DBT concept called healthy self-doubt
Mindfulness of consumption	DBT mindfulness skills; gratitude practice

Kaza (2008).

of their green practice. Also, she describes two types of seekers: Systematic and serendipitous. While systematic seekers create specific plans with action items, the serendipitous seeker does not have an ironed-out plan but has faith that life will present opportunities that will lead them to their path.

CBT correlates

1. *Values.* A foundational theme of ACT (Twohig et al., 2020) is that there is divergence between one's values, or how we aspire to live, and the extent to which your behaviors are in line with those values. Reflecting on the things you value and how you express your values can help you create action items, as in systematically seeking goals that will allow you to live in accordance with your values.
2. *Radical Openness.* RO DBT (Lynch, 2018) is a cognitive-behavioral treatment that specifically targets maladaptive behaviors resulting from overcontrolled personality traits, such as extreme perfectionism, flat emotional expressions, and trouble developing intimate relationships. Not everyone with an overcontrolled personality has maladaptive behaviors, and in fact, overcontrol is considered prosocial in encouraging responsible, productive behavior. However, the ambition of overcontrolled individuals to work harder might hamper their ability to nurture social connections.

Radical Openness (Lynch, 2018) is a core principle in RO DBT that refers to being open to learning from others even when they provide *disconfirming feedback*, meaning information that differs from what you believe is right. This is similar to what Kaza (2008) describes as serendipitous seeking, with both being characterized by what RO practitioners refer to as *flexible control*, the state of being open to changing without giving in to others in a defeatist way.

3. *RO DBT ambassador principle.* The role of the RO DBT therapist is to guide the client in rejoining their social circle. The role of the therapist metaphorically is one of an ambassador in encouraging the socially isolated OC client to rejoin their social circle.
4. *Compulsive behaviors.* Compulsive behaviors are conceptualized by RO DBT practitioners as an overcontrolled coping mechanism, whether this be compulsive fixing or correcting, apologizing excessively, or avoidance. Generally, there is an urge and accompanying state of discomfort from which compulsive behavior follows. This is similar to what Kaza (2008) describes as *shenpa* or the natural discomfort with a world that is changing and which is accompanied by urges to seek things that bring us comfort. Kaza recommends identifying *shenpa* when it arises and resisting urges to cave into it (this resistance is known as *shenlok*). This can be useful in curbing environmentally damaging practices that waste resources. By learning to sit with discomfort, we become more resilient and help the earth.
5. *Healthy self-doubt.* Healthy self-doubt is an RO DBT concept that refers to a temporary state of mind beginning with the assumption that we don't see things as they are but instead see things as we are (Lynch, 2018). It is pausing to consider that our thoughts and behaviors might be inaccurate or ineffective without harshly criticizing ourselves or blaming others. Healthy self-doubt is similar, at least in part, to Kaza's (2008) description of the *Green Lifeway*, which focuses not so much on the specific decisions we make (i.e. recycling or purchasing an electric car) as the intention or spirit behind our way of living. This includes recognizing when we are not acting in a way that is kind to the earth or mindful of its limited resources. This idea ties into her concept of mindfulness of consumption, further explored below.
6. *Mindfulness to consumption.* The positive effects of mindfulness on mental health have strong support in the research literature. The way mindfulness works is that our capacity for focused attention is limited, and when we commit to focusing on minute details of the present moment there is less mental energy available for mood sappers such as rumination, perseveration, and judgmental thinking about

oneself and others. However, DBT Mindfulness skills (Linehan, 1993) also instruct clients to be aware of unhealthy thoughts without judging yourself for judging yourself, but simply allowing negative thoughts to pass through the mind like clouds floating across the sky.

Similarly, we can attend to our consumption without responding with self-deprecation. For example, by noticing that I took a longer shower than necessary, I increase my awareness of my current water consumption, and this makes it more likely that I will remember my resolve to save resources the next time I shower (and ultimately to reduce my water usage).

Working toward change is far more effective and rewarding than reprimanding yourself. Remembering the cycles inherent in nature reminds us more opportunities will follow to act in accordance with our values.

Before we leave this section, let's take a moment to discuss an ongoing controversy in the climate psychology world: whether it is the role of mental health professionals to inspire their clients to engage in pro-environmental behavior. As mentioned earlier in this book, research favors this approach (Schwartz et al., 2023) particularly via collective action, like joining a pro-environmental groups the Citizens' Climate Lobby (CCL) or the Sunshine Movement. This makes sense given the opportunities for social connection.

However, I would also recommend starting with encouraging exploration of the client's interests and skills. For instance, Kaza (2008) describes how some get involved in the cause out of ambition, and this means that skills will be more important for the client to derive meaning from their actions.

I would also recommend coupling any encouragement of action with a reflective or spiritual activity, such as spending time in nature or gratitude journaling. Such reflective activities serve the purpose of reminding the individual why they are doing this work. This is particularly important for preventing burnout in climate work.

Thus, spirituality both catalyzes and sustains climate action. At the same time, there are dangers to relying on it as the end all be all to managing eco-distress. I point out these caveats below.

STEERING CLEAR OF BLIND FAITH

In his essay *The Guilted Age of Carbon* (2023), Tad DeLay critically examines not only the ideology of climate deniers but also climate advocates who allay guilt through practices like carbon offsetting, such as building trees to compensate for burning fossil fuels.

DeLay talks about two types of denial. One is what most of us think of in relation to climate denial or denying the existence of human-caused climate change.

The other form of denial is what he refers to as the guilt denial of liberalism. This is the unrealistic belief that we can continue to live exactly the way we have come to live in modern times, and simultaneously do things to save the planet. The reality is that most of us are collectively ignorant of how many vegetarian dinners will make up for the flight to Hawaii. As DeLay puts it,

> .... Afforestation is indeed an important tool in emissions reduction – but a deeply limited one. A mature tree absorbs 22 kilograms of carbon dioxide per year. An average American lifestyle annually emits 14.5 metric tons of CO2. Were I to scrub those emissions .... I would need to plant about 660 trees.... (para. 13)

At the same time, DeLay (2023) has two children and talks about the human instinct to hope for a future. His encouragement of supporting politicians who are relatively pro-environmental, and supporting climate activists in any capacity suggests that even if individuals have limited agency in solving the climate crisis, our spirit should be pro-conservation and environmental, nonetheless.

DeLay's (2023) stance provides yet another rationale for cultivating an ability to understand dialectics (two opposing ideas that when synthesized resolve the tension they pose – Merriam Webster online dictionary, 2025) and sit with the tension that arises from them. I know that driving an electric car will not be enough to help the climate crisis, but I do it anyway because it is enjoyable to me (not having to go to a gas station has been great), and it is in line with my values of environmentalism. Even though electric cars are more environmentally friendly, I am aware that our decision to buy a second vehicle was harmful to the environment. On the other hand, I have taken up practices like biking and timing errands to limit needing to use the vehicle. Holding these truths simultaneously is a dialectic.

DeLay's arguments for the lifestyle shifts that are necessary to sufficiently address the climate crisis are consistent with the concept of permaculture (e.g. Reiff et al., 2024). A portmanteau, or contraction, of the words "permanent" and "agriculture," permaculture is the philosophy of working with rather than against nature. It entails restructuring the fundamental systems (e.g. agricultural, technological, economical) that usurp nature into systems that are in line with natural systems and protect the earth. Another piece of the philosophy behind permaculture is rehabilitation (for instance, of deforested areas, or by creating refuges for endangered species).

Peter Kalmus (2017), author of *Being the Change: Live Well and Spark a Climate Revolution* (chapters are available for free through his website at https://peterkalmus.net/books/) and a climate scientist known for his vociferous social media posts and bold activism, discusses opting out of a broken system that dictates how we use energy and natural resources.

Ideas Kalmus presents for opting out include: Opting out of stuff (i.e. material possessions, both purchasing less and getting rid of it); repairing instead of buying new and repurposing and using low-energy technology like push lawn mowers, solar ovens, treadle sewing machines and bicycles. He also recommends supporting local economies; and practicing freeganism, or using food or other items that others have thrown away. Freeganism can include dumpster diving, although there are programs that allow people in the community to obtain discounted restaurant leftovers without having to resort to foraging a dumpster. It can also include minimizing packaging by making your own food or being creative with leftovers to avoid throwing out food.

CONCLUSION

Spirituality is distinct from religion and, in the context of environmentalism, can intertwine environmentalism with other values and passions. For example, I usually bake my own bread for my family, which saves emissions from transporting bread to a store and saves packaging. I had originally thought of baking, though, as a way to get my children to eat healthier foods that aren't loaded with preservatives.

Then, a friend of mine noted a truth I couldn't deny, that baking was also meditative. It forces me to take my mind off the horrors of current events and the stress of tasks yet to be completed because, as anyone who bakes knows, concentrating on the measurements and techniques outlined in a recipe is essential for creating treats that are tasty, or at least edible.

Spirituality in climate change serves an important purpose in increasing willingness to turn toward the pain of eco-distress, but it is far from the final solution. Spirituality has the power to get the ball rolling, but if we lean too much into the notion that climate change involves mystical powers beyond our control, then we will fail to take adequate action to push our governing bodies to do what we can't do alone: Stop the fossil fuel industry. Nevertheless, spirituality will allow us to authentically get behind this mammoth-like but essential objective for fighting climate change.

REFERENCES

Ahtuangaruak, R., (November 24, 2022). Don't make indigenous people pay Willow's price. *The Hill.* https://thehill.com/opinion/energy-environment/3749037-dont-make-indigenous-people-pay-willows-price/

Allende, I. (1989/1991). And of clay are we created (M. Sayers Peden, Trans.) In A. Charters (Ed.), *The story and its writer: An introduction to short fiction* (pp. 30–37). Bedford/St. Martin's.

Asian Research Center for Religion and Social Communication Administration (2023, January 17). *Environmentalism for the post-pandemic world: Lessons from Jainism and Jain monks.* https://asianresearchcenter.org/blog/essays/environmentalism-for-the-post-pandemic-world-lessons-from-jainism-and-jain-monks#:~:text=This%20ancient%20Jain%20principle%20teaches,needs%20as%20far%20as%20possible

Beard, A. (May 2016). Life's work: An interview with Isabel Allende. Creativity. *Harvard Business Review* (from the Magazine). https://hbr.org/2016/05/isabel-allende

Bell, C. N., St George, R., Honan, C., & Matthews, A. (2025). From forest to focus: The interactive effects of nature exposure and nature relatedness on attention, brain activity, heart rate variability, and mood. *Journal of Cognitive Neurosicence, 38*(5), 1–23. https://doi.org/10.1162/JOCN.a.2416

Billet, M. I., Baimel, A., Sahakari, S., Schaller, M., & Norenzayan, A. (2023). Ecospirituality: The psychology of moral concern for nature. *Journal of Environmental Psychology, 87,* 102001. https://doi.org/10.1016/j.jenvp.2023.102001

Cruz, G. (2025, April 26). Isabel Allende understands how fear changes a society. The Interview. *The New York Times.* https://www.nytimes.com/2025/04/26/magazine/isabel-allende-interview.html?login=google&auth=login-google

Dayenu: A Jewish call to climate action. (2026). *Who we are.* https://dayenu.org/who-we-are/

DeLay, T. (2023, July 5). The guilted age of carbon. Criticism & Essays. *Protean Magazine.* https://proteanmag.com/2023/07/05/the-guilted-age-of-carbon/

Dias, E., & Graham, R. (2022, April 6). The growing religious fervor in the American right: 'This is a Jesus movement.' *The New York Times.* https://www.nytimes.com/2022/04/06/us/christian-right-wing-politics.html

Dopko, R. L., Capaldi, C. A., & Zelenski, J. M. (2019). The psychological and social benefits of a nature experience for children: A preliminary investigation. *Journal of Environmental Psychology, 63,* 134–138. https://doi.org/10.1016/j.jenvp.2019.05.002

Doyne, S. (2024). Are teenagers obsessing too much about skin care? Student Opinion. *The New York Times.* https://www.nytimes.com/2024/03/25/learning/are-teenagers-obsessing-too-much-about-skin-care.html

Dreyfus, J. (2023, April 11). Julia gets wise with Isabel Allende. *Wiser than me.* Lemonada Media.

Dreyfus, J. (2024, May 15). Julia gets wise with Anne Lamott. *Wiser than me.* Lemonada Media.

Earth Care Congregations. (2025). *A guide to greening presbyterian churches (version 5).* https://pcusa.org/resource/earth-care-congregations-guide-greening-churches#:~:text=Version%205%2C%20updated%202025,Guide%20to%20Greening%20Presbyterian%20Churches.%E2%80%9D

Frenken, M., & Imhoff, R. (2025). Law and order, magical spirits, and false flag operations: On the co-occurrence of authoritarianism, spirituality, and conspiracy beliefs and their association with support of violent protests. *Political Psychology, 47*(1), 1–21.

Goldman, L. (2022). *Climate change and youth: Turning grief and anxiety into activism.* Routledge.

Hassan, A. (2019). 'Thoughts and prayers' aren't enough, America's first gun violence minister says. *The New York Times.* https://www.nytimes.com/2019/07/28/us/minister-gun-violence-church.html

Hoffman, D. (2017, December 1). *The Tulane Hullabaloo.* https://tulanehullabaloo.com/34291/arcade/isabel-allende/

Irvine, W. (2019). *The stoic challenge: A philosopher's guide to becoming tougher, calmer, and more resilient.* W.W. Norton & Company.

Jesko, J. (Director), Williams, R. R., Waterfield, A., Rodriguez, S., Heller, L., Capote, N., & Abraham, N. (Executive Producers). (2023). *Savior complex* [TV mini-series]. Latchkey Films; One Story Up Productions; HBO Documentary Films.

Kalmus, P. (2017). *Being the change: Live well and spark a climate revolution.* New Society Publishers.

Kaza, S. (2008). *Mindfully green: A personal and spiritual guide to whole earth thinking.* Shambhala.

Linehan, M. (1993). *Cognitive-behavioral treatment of borderline personality disorder.* The Guilford Press.

Luetz, J. M., & Nunn, P. D. (2023). Spirituality and sustainable development: An entangled and neglected relationship. *Sustainability Science, 18,* 2035–2042. https://doi.org/10.1007/s11625-023-01347-8

Lynch, T. (2018). *Radically open dialectical behavior therapy: Theory and practice for treating disorders of overcontrol.* Context Press.

Merriam-Webster. (2025). Dialectics. In *Merriam-Webster.com dictionary.* Retrieved June 24, 2025, from https://www.merriam-webster.com/dictionary/dialectic

Nisbet, E. K., Zelenski, J., & Grandpierre, Z. (2019). Mindfulness in nature enhances connectedness and mood. *Ecopsychology, 11*(2), 81–91. https://doi.org/10.1089/eco.2018.0061

Presbyterians for Earth Care: An Eco-Justice Network. (2020). *Introduction to Earth Care Congregations.* https://presbyearthcare.org/earth-care-congregations-webinar/

Quyet, N. T., Lan, P. T., & Phuong, N. T. (2022). The four noble truths an integrative Buddhist philosophy of life. *European Journal of Science and Theology, 18*(6), 47–63.

Reiff, J., Jungkunst, H. F., Mauser, K. M., Kampel, S., Regending, S., Rösch, V., Zaller, J. G., & Entling, M. H. (2024). Permaculture enhances carbon stocks, soil quality and biodiversity in Central Europe. *Communications Earth & Environment, 5,* 305. https://doi.org/10.1038/s43247-024-01405-8

Schwartz, S. E. O., Benoit, L., Clayton, S., Parnes, M. F., Swenson, L., & Lowe, S. R. (2023). Climate change anxiety and mental health: Environmental activism as buffer. *Current Psychology, 42,* 16708–16721.

Sheffield, D., Butler, C. W., & Richardson, M. (2022). Improving nature connectedness in adults: A meta-analysis review and agenda. *Sustainability, 14*(19), 12494. https://doi.org/10.3390/su141912494

Singh, R. P. B., Rana, P. S., & Olsen, (2021). The environment as a sacred space: Religious and spiritual tourism and environmental concern in Hinduism. In *The Routledge handbook of Religious and spiritual tourism.* Routledge.

Stein, A. T., Carl, E., Cuijpers, P., Karyotaki, E., & Smits, J. A. J. (2021) Looking beyond depression: A meta-analysis of the effect of behavioral activation on depression, anxiety, and activation. *Psychological Medicine, 51*(9):1491–1504. https://doi.org/10.1017/S0033291720000239

Stone, L. M. (2023). *And of clay are we created by Isabel Allende.* EBSCO. https://www.ebsco.com/research-starters/literature-and-writing/and-clay-are-we-created-isabel-allende

Suganthi, L. (2019). Ecospirituality: A scale to measure an individual's reverential respect for the environment. *Ecopsychology, 11*(2), 110–122. https://doi.org/10.1089/eco.2018.0065

Tejeda, M. J. (2025). Ecospirituality, prosumerism and the nature-centered practices in modern witchcraft. *Journal of Management, Spirituality & Religion.* https://doi.org/10.51327/HSHP5579

Twohig, M. P., Levin, M. E., & Ong, C. W. (2020). *ACT in steps.* Oxford UP.

Van Kessel, P., Smith, G. A., & Schiller, A. (2018, November 20). *Where Americans find meaning in life.* Pew Research Center. https://www.pewresearch.org/religion/2018/11/20/where-americans-find-meaning-in-life/

Chapter 6

Radically Open Dialectical Behavior Therapy (RO DBT) for Eco-Perfectionism

RATIONALE FOR TARGETING PERFECTIONISM IN ECO-DISTRESS

One morning, I woke up feeling exhausted after a poor night of sleep but pushed myself to get started on the morning routine of preparing lunches and ironing clothes. I tried to pull socks out of the dryer only to realize it was empty, and that an entire load had been sitting in the washing machine for the past two days. Opening the washing machine, my heart sank as the stench of wet clothes hit my nostrils.

If you've found yourself in a similar situation, what kinds of thoughts and emotions did you have?

It would make sense to feel a negative emotion given that the expectations for fresh, clean, clothes were met with the reality of a stinking mess. You might feel annoyed at having to take the time and effort to redo the job or going another few hours without the socks you were hoping to wear.

If you are someone who struggles with perfectionism, though, you might have trouble moving on from the fact alone that you made a mistake. And if you are someone who struggles with *eco-perfectionism*, meaning your intolerance of mistakes is focused on mistakes that result in environmental harm (however small this may be in the scheme of things), the lapse might trigger intense anxiety and self-criticism. Conventional washing machines use about 40 gallons of water per load (Portland.gov, 2018–2025). Water scarcity is one of the scariest aspects of the climate crisis.

On the surface, this vignette might look like garden-variety eco-distress, but it's important to also consider how overcontrolled personality traits,

DOI: 10.4324/9781003489665-7

including perfectionism, factor into it. While trying to avoid mistakes might sound like a good thing, living with intense fear of them isn't. For one it's counterproductive, as research shows that people who try to avoid mistakes end up making more (Edmondson, 2023).

Second, perfectionism fuels excessive self-criticism. Extreme thinking like "I'm an idiot" feeds off the perfectionistic paradigm in which mistakes are not to be tolerated, even minor ones (Curran, 2023). Contrary to popular opinion that being a perfectionist is fodder for humble bragging, it is actually a cause for concern.

Perfectionism expert Thomas Curran (2023) emphasizes this point in his seminal work, *The Perfection Trap.* As I've mentioned in previous chapters, Curran and Hill's (2019) research has shown that since the 1980s, perfectionism has been on the rise. Curran (2023) discusses the adverse consequences of perfectionism to mental and physical health, as well as to the environment via the effect of unabated consumerism on exacerbating climate change. According to Curran, the core of perfectionism is the belief of not being good enough. It is fueled by an unremitting idea that you have a pervasive deficit that must be overcome or fixed through willpower or external resources, such as drugs and material possessions. For example, if we believe ourselves to not be attractive enough, we must buy more beauty products to look better. If we aren't smart enough, we burn ourselves out relentlessly striving to fill the void.

Throughout this chapter, I refer to perfectionism as maladaptive perfectionism, but keep in mind that according to Curran (2023), no level of perfectionism is healthy. In driving one to cover up perceived deficits, the perfectionist reacts to minor errors or inconsistencies with defensive arousal of the sympathetic nervous system, the main player in the well-known "fight or flight" response (Lynch, 2018a). One might respond to this uncomfortable arousal in various ways, including sulking in self-blame and criticism, which could lead to avoidance, giving up, compulsive fixing or over-correction. This aligns with the concept of *self-oriented perfectionism* (Curran & Hill, 2019).

Relatedly, *socially prescribed perfectionists* are hypervigilant to social judgment and tend to assume that others are expecting flawless behaviors (Curran & Hill, 2019). Therefore, we will be seen as failures in response to any deviation from perfection. Socially prescribed perfectionism may be accompanied by fear of being rejected by others and frustration about the world being an unforgiving place.

Yet another response to the defensive arousal triggered by perfectionism is to blame others for their imperfections, a tendency that fits with *other-oriented perfectionism*, or having unrealistic expectations of others.

HOW DO DIFFERENT TYPES OF PERFECTIONISM RELATE TO ECO-DISTRESS?

Consider the following: Recently, I came across an online post from an environmental activist advocating skipping a daily shower. My immediate response was indignation. How about corporations go without excessive profit at the cost of dumping chemicals into rivers and lakes, or people focus on purchasing fewer gadgets before we go after hygiene?

RO DBT, as explained below, guides me in pausing and reflecting on this defensiveness and attempting to open myself up to other opinions without blindly agreeing with them. While I still might not agree with the post, this self-reflection might help me understand that the negativity I felt might at least in part be due to it triggering a belief in "not being good enough." Given that humans need to consume natural resources in order to survive, perfectionists who are aware of environmental problems might be more prone to never feeling like they are being good enough environmentalists.

Perfectionism, though, hurts more than self-esteem alone. It also does a number on relationships. For example, a defensive response to the laundry situation above might be, *Why isn't anyone else helping me?* This might be an example of other-oriented perfectionism if the person did not ask anyone to help, and is likely to brew resentment, a known intimacy-killer. Self-oriented perfectionism, on the other hand, can make someone too self-focused, as it requires a lot of time and attention. This, in turn, makes it hard to prioritize relationship-enhancing activities, which might come to be viewed as a waste of time. Socially prescribed perfectionism goes hand in hand with excessive people pleasing, which can lead to being too sensitive to criticism from say, a leader of an environmental group you have joined.

The cruel irony of perfectionism is that, in some cases, it leads to lost productivity, such as when someone procrastinates out of fear of failure. But even when perfectionists are high achieving, their relentless pursuit of flawlessness comes at the expense of creativity and challenging themselves, which they perceive as being too risky. Most pertinent to the psychology of climate change though is the lost opportunity for social connection, which could limit one's potential to engage in collective climate action, such as being able to work well with others in an environmental group.

Examples of where perfectionism can get in an environmentalist's own way might be chastising people for running the faucet a few seconds longer than needed (other-oriented perfectionism), which can be a counterproductive overreaction that can make people think of this person as unkind, mentally unwell, or both. Unfortunately, they might feel less positively about that person's cause (i.e. climate change) as well.

Another example of this would be refusing to accept energy-wasting gifts (self or other-oriented perfectionism) even though it brings a loved one joy to give them. Could alternate options be considered, such as making sure the item is used well and ultimately repurposed? Such questions help foster the kind of flexible thinking and problem-solving that is at the heart of sustainability without compromising treasured relationships.

WHAT IS RADICALLY OPEN DIALECTICAL BEHAVIOR THERAPY (RO DBT) AND HOW DOES IT HELP PEOPLE WITH ECO-PERFECTIONISM?

Radically open dialectical behavior therapy (RO DBT) is an evidence-based, cognitive-behavioral treatment for psychological disorders of overcontrol (Lynch, 2018a). Overcontrolled personality traits such as being organized and ambitious are part of the typical spectrum of behavioral traits and are largely prosocial, meaning that people with overcontrolled personalities are vital to a thriving social order. These people get stuff done. They wake up early, work overtime, and relentlessly pursue productivity, accuracy, and progress. They are good at delaying gratification and prioritizing work over play.

Overcontrolled traits are ideal for creating a more sustainable future, which will require persistence and attention to detail. But this gift can come with side effects and not being great at having fun means perfectionists are prone to feeling disconnected from others, as well as more prone to loneliness and feeling misunderstood (Lynch, 2018a). RO DBT aims to create a life worth sharing by helping people understand how their social behaviors are unintentionally pushing others away and teaching them skills that are more conducive to feeling connected.

Although we don't yet have research exploring the effectiveness of RO DBT in supporting engagement in global justice issues, there is evidence that a sense of belongingness to one's community increases the likelihood of engaging in civic action (Over Zero and the American Immigration Council, 2024). Therefore, while being concerned about climate change and wanting to do something about it are necessary components of being a social changemaker, one also needs the tools and opportunities to affect larger systems that are driving environmental devastation. Individual-level pro environmental actions such as taking shorter showers are admirable choices, but having a broader effect on the community is facilitated by feeling like you belong in it to begin with.

RO DBT is a unique intervention in the sense that it doesn't focus on the internal triggers of emotional problems, such as maladaptive beliefs and emotional intensity. Rather, it aims to address specific external behaviors known as *social signals* (Lynch, 2018a).

Social signals refer to behaviors that are done in the presence of another person (Lynch, 2018a). People with excessive overcontrol can have social signals that unintentionally push others away. These include verbal behaviors, such as excessive apologizing, and nonverbal behaviors such as flat facial expressions or, in contrast, excessive smiling. RO DBT focuses on these kinds of behaviors because excessively overcontrolled individuals tend to lack congruence between their intention and social signaling. For instance, they might smile and nod excessively even though they disagree with the other person or display a flat expression and a monotone even though they are interested in what the other person has to say. This lack of congruence is part of what leads to them feeling misunderstood and disconnected with others.

People with excessive overcontrol are also hesitant to pursue novel situations (Lynch, 2018a). Two summers ago, I moved from a population-dense town without sidewalks to a town that had bike paths and less traffic. The idea of riding my bicycle to work, which would have been aversive where I lived previously, became tempting given its potential to reduce my carbon footprint, as well as to save money on parking fees and get a great workout. However, my overcontrol made me hesitant. I worried about looking strange or getting lost. Then I realized that many others bike to campus. And while I got lost at first, with a little practice, I had no problem. This experience exemplifies the idea that that openness to novel experiences can lead an improved quality of life.

The evolution of my turning from someone who was averse to biking to work to someone who can't get enough of it, rain or shine, represents an intriguing analogy to the process we will inevitably have to undergo in altering our lifestyles to become more sustainable. Given that I hadn't ridden a bike in over 20 years, the idea felt crazy and scary. "What if" thoughts flooded my brain, and my instinctual response was to fill my Amazon cart with every bike gadget to make the ride more comfortable. Initially, my list of conditions for biking was so long that it left only a small sliver of days for which it would be feasible (i.e. sunny weather, not too hot, not too cold, must be dry, must not have any late meetings). However, with experience and proper planning, I adapted to biking in a broader range of conditions. I enjoyed biking so much that I found ways to do it in the rain, on 80 degree days, in the dark, etc. In other words, I made adaptations, such as repurposing old elastic pants I would otherwise have discarded by wearing them over my work attire to avoid getting mud on my nicer clothes.

It is worth mentioning that RO DBT helps people shape their behaviors without dramatically altering their overcontrolled personalities, because as I mentioned earlier, such personalities are to a certain extent, prosocial. An example of a simple skill that can reshape the impression you give off to others without altering the core of your personality is raising your eyebrows (known as an "eyebrow wag"), which signals openness to others. Others positively respond in milliseconds in what RO DBT refers to as a neurobiological social safety system (Lynch, 2018b).

Another target of RO DBT is identifying and reshaping maladaptive coping styles, such as when avoidance is used excessively. For example, avoidance in moderation, such as taking breaks from reading the news, including discouraging news about environmental policy, can be helpful to mental health and burnout prevention, while refusing to acknowledge problems like climate change is due to excessive avoidance that deprives someone of taking part in a solution.

Another target of treatment is addressing overcontrolled coping styles. Approach coping represents a proactive attempt to solve problems, such as coping with the anxiety of having a project to complete by getting it done, or coping with eco-anxiety by joining an environmental group. However, in problems of overcontrol, approach coping can take the form of compulsive fixing. Over-fixing can lead to missing the forest for the trees as well as lead you down the fast track to burnout, particularly when combined with detail-focused processing, selective attention toward perceived errors, and magnification of these errors, all of which are other potential problems that are common in people with overcontrolled personalities.

As a hypothetical example, take an activist who has the opportunity to discuss environmental legislation with a government official, but misses their appointment because they refuse to drive there. Or refusing to join an environmental advocacy group because you noticed how many of the members brought disposable coffee cups to the monthly meeting.

With climate change, we can see these problems playing out on a broader social scale. People are afraid to adapt. They don't know how they will be able to live differently, just like I was scared because I didn't know what riding a bike to and from work would be like. Would people laugh at me? Would I arrive at the class I'm teaching that day dripping in sweat? All those things happened, but I took them in stride because I enjoy biking so much and believe in its benefits both for my health and the health of the planet.

To be successful with a novel endeavor in the long run, one needs to be able to cope with failures, seeing them in a realistic light as opposed to an exaggeration in their mind. This means that those who are hesitant to live more sustainably might need to cope with mild discomfort that will lead to pay-off down the line. It also means that environmentalists will need to cope with imperfection.

Transitions are invariably hard. Mental health practitioners can help people adapt by normalizing and validating emotions, and by sharing tools that help them better lean on others for support.

HOW DOES RO DBT WORK?

RO DBT is an evidence-based intervention (Lynch, 2018a), supported by research for treating disorders such as obsessive compulsive personality disorder, treatment resistant depression and anxiety disorders, autism spectrum disorder (in individuals with low support needs, previously referred to as high-functioning autism) and anorexia nervosa (Lynch, 2018a).

The theoretical basis for RO DBT is described as follows: (1) we are a tribal species that evolved to form lasting, trusting bonds with others, and (2) psychological health is rooted in the confluence of three features: Openness, flexibility, and social connectedness. At treatment outcome, the aim is for the individual to see relief in psychological symptoms associated with emotional distress, such as depression and anxiety by having more satisfying social connections.

RO DBT therapists help their clients identify social signals that are blocking them from connecting with others, such as helping them modify nonverbal behaviors to express openness and curiosity, and encouraging a level of emotional expression that matches how they are actually feeling. By encouraging clients to share uncomfortable emotions, such as embarrassment with others rather than try to hide or correct them, they come to see that such self-disclosures help them feel more accepted and likeable.

In the process of adjusting social signals, the client is guided in understanding their "valued goals" for social connection. Then, they are guided in understanding maladaptive coping behaviors that are interfering with living in accordance with these goals.

Take, for instance, someone who desperately wants to be seen by others as competent. They go about trying to achieve this by avoiding mistakes. They correct every slight grammatical error, apologize excessively for inaccuracies that are so minor that others either wouldn't have noticed or cared, and don't express their ideas or opinions out of fear that others will see flaws.

Ironically, this person comes to be undervalued for their intelligence and might even risk coming off as less competent or emotionally unstable. Therefore, the individual in question might be guided in practicing stating their mind with more confidence, and to inhibit excessive apologizing. In this way, they learn to act in ways that are more congruent with their desire to be seen as competent by others.

Importantly, the individual is guided in understanding that behaviors such as over-apologizing for minor errors and excessively inhibiting sharing ideas is their way of coping, and that this way of coping can be beneficial in certain situations, such as when they have offended other people, while being self-defeating in others, such as when they are trying to stand by their decisions to engage in sustainable behavior unapologetically.

While RO DBT focuses on external, social behavior, it doesn't ignore the need to address inner turbulence that is fueling maladaptive social signaling. Such internal thinking patterns are addressed in a lesson called *Overcontrolled States of Mind* (Lynch, 2018b), which I apply below to the laundry situation I described at the beginning of the chapter.

Returning to the example of my laundry fiasco, one can see a vacillation between cognitive extremes, from self-blame for letting the clothes get stinky, to blaming others from expecting other people in the house to pick up my slack without having asked for that kind of help.

In RO DBT, the terms that describe extreme and unhealthy states of mind are defined as *fixed* and *fatalistic* (Lynch, 2018b). RO clinicians assist clients in identifying these states of mind and guiding them through specific skills that help them lean into helpful *Flexible-Mind*. At the same time, RO emphasizes that it is neither realistic to expect to be in a flexible state of mind all the time, nor is it healthy. For example, an absolute "no" isn't being rigid if someone is trying to get you to do something dangerous or illegal.

Fixed mind operates on the premise that "Change is unnecessary because I already know the answer" (Lynch, 2018b, p. 245). Thomas Lynch, who developed RO DBT, uses the analogy of the Titanic's captain, whose stubborn refusal to slow down the speed of the ship to heed ice bergs captures this perspective. As mentioned above, RO DBT emphasizes that there are contexts for which Fixed Mind is appropriate. For example, fixed mind would make sense if someone were trying to convince you that the world were flat (yes, there are people who still believe it is), and it would be adaptive in allowing you to stay true to your knowledge and convictions, such as standing behind the data showing that climate change is real and caused by human activity. In other words, being flexible does NOT mean blindly agreeing, and psychological flexibility does not mean perfect flexibility, which would defeat the purpose of it.

In contrast to Fixed Mind, *Fatalistic* Mind is known as the mental escape artist (Lynch, 2018b), operating on the premise that "change is unnecessary because there is no answer" (p. 245). Lynch uses the example of the Titanic's captain secluding himself rather than trying to help crew mates and passengers. Fatalistic Mind underlies problematic ways of seeing the world, such as from a victim mentality (e.g. "everyone is out to get me"), excessive self-criticism ("I'm a hopeless case"), and nihilism ("What's the point. Life is meaningless"). What is implicated in all of these problematic mindsets is the unwillingness to take responsibility for one's own emotions, beliefs, and behaviors, including mistakes. One can see this being a vulnerability to the eco-distressed given the enormity of the climate crisis (i.e. thinking "What's the use of trying? Clearly the world is screwed!"). It is also thematic of Elon Musk's mission to inhabit

Mars, rather than address the problems faced by the one planet we know for certain can support human life.

RO DBT again brings context into consideration, by highlighting that Fatalistic Mind could serve as an important sign that something is wrong with our current situation, or that we need to rest or consider shifting direction. An example of shifting from Fatalistic Mind to Flexible-Mind is catching the thought, "I need to quit trying to be sustainable" and considering whether this is a sign that a less drastic shift is more effective. As an example, my family and I purchased solar panels for our previous home. We took a break from using solar panels when we moved and our new Home Owners Association (HOA) prohibited solar panels. Recently, our HOA announced that they would be permitting solar panels, so we looked into having them installed within the next few months. Unfortunately, financial incentives the government previously offered have been revoked, and solar panels have become prohibitively expensive. At the time of writing this, then, they remain on my wish list for our current home.

Consideration of overcontrolled states of mind can be applied to eco-distress in several ways. Constant doomscrolling, which clients have shared they were bullied into doing to prove that their empathy and activism were genuine enough, can sabotage engagement efforts by fueling fatalism. This example also portrays how fatalism can fuel divisiveness, which is emotionally harmful and counterproductive given the importance of cohesive, collective efforts in climate action.

The question of whether or not to have children is another example of how Flexible Mind can help us appreciate the importance of context. Although children use considerable resources and come with immense responsibility, they also represent beacons of hope. The decision to not have children, in and of itself, is not fatalistic. But to ensure unity, it's also important to understand why environmental advocates might want to have children of their own.

As mentioned earlier, some young adults are deciding to forgo having children due to the uncertain state of the world (Wray, 2022). I respect this decision from the perspective of women's right to autonomy over their bodies. At the same time, I am cautious that in some cases it may speak to the mindset that all hope is lost (in other words, a fatalistic one).

As described earlier, extreme prepping, such as building bunkers without trying to mitigate the effects of climate change, is another example of Fatalistic Mind. The individual is attempting to run away from the problem without trying to contribute to the solution.

Fatalistic Mind is something I have experienced when I perceived activists looking down on me for not knowing enough about heat pumps or criticizing me for not being bold and savvy enough to convince others to join our cause. These moments led me to lose a sense of meaning and to feel like giving up.

I came to realize, though, that the problem with adopting a victim mentality is that nothing gets done. Luckily, I've shifted away from Fatalistic Mind by continuing to explore how to best use the skills I have that could be helpful. Writing turned out to be the best outlet for me.

On the other hand, pressuring people to doomscroll might stem from another problematic state of mind: Fixed mind, explored below. Below is a portrayal of overcontrolled states of mind, adapted for eco-distress. In this case example, denialism corresponds with Fixed Mind, doomer thinking corresponds with Fatalistic Mind, and adaptive thinking corresponds with Flexible-Mind, representing proactive responses to eco-distress.

CASE STUDY

Sandy, an organizer, is frustrated with her volunteers, who are manning a table for their environmental advocacy group at a local festival. One volunteer went into too much detail about solar panels, while another had trouble remembering the facts of an energy bill that the group was promoting. Yet another asked for a lunch break even though they hadn't recruited a single new volunteer.

Sandy's therapist presents a handout from RO DBT Skills class, a 30-week group focused on practicing RO DBT skills meant to be completed in conjunction with individual treatment. The handout is called *Being Kind to Fixed Mind* (Lynch, 2018b, pp. 250–253). Below, I apply these skills with the case of Sandy, which, like other case studies highlighted in this book, represents a conglomeration of situations similar to ones faced by clients. We will first apply skills for addressing fixed mind. We then apply skills for fatalistic mind, from the handout *Going Opposite to Fatalistic Mind* (pp. 252–253) to a different case with a hypothetical client named Darren.

From *Being Kind to Fixed Mind* (Lynch, 2018b, pp. 250–251).

Step 1. When Challenged, Observe Emotions, Action Urges, and Thoughts That Might Be Linked to Fixed Mind

Sandy acknowledges that she feels frustrated and thinks that her colleagues are misguided and not committed.

Step 2. Acknowledge the Possibility of Being in Fixed Mind

Sandy is guided in the following:

1. *Finding physical tension in her body, which can be a sign of Fixed Mind.* Sandy expresses feeling hot and tight in her stomach while managing her group's table.
2. *Use self-enquiry to facilitate self-awareness.* Self-enquiry is a pivotal skill across RO interventions in which the client reflects on thoughts and feelings and is urged not to jump to solutions but rather find a good question that will get them closer to their "edge," or actions, thoughts, feelings, images or sensations that we usually try to avoid because they reveal personal vulnerabilities such as fear of criticism. Sandy asks herself, "What is there to learn here? What am I afraid might happen if I were to momentarily drop my perspective [that her colleagues are misguided and not committed]?"
3. *Acknowledge that you are fighting or resisting something without mindlessly letting go of your point of view.* The point is not to avoid avoidance but rather to better understand urges to avoid. Sandy acknowledges that she has urges to avoid considering her colleagues' perspectives.
4. *Gently remember that when in Fixed Mind, your thoughts, emotions, urges, and sensations are determined by your past learning.* Sandy grew up in a home where mistakes were deemed intolerable and idle time such as watching television during the day was harshly criticized. Her parents would call her lazy if she left her room untidy or called on friends before chores and homework had been completed. She enjoyed art but her parents pointed out minor flaws in her drawings. Her mother was an artist who insisted that her technique be "just right." Sandy acknowledges that, influenced by these past experiences, she tends to be excessively critical toward herself and others.
5. *Remind yourself that Fixed Mind alerts us to those things in our life we need to be more open to in order to improve ourselves or learn more.* Sandy doesn't really want to unleash her wrath on her helpers. She needs them, and they are competent and hard-working. Her therapist works with her to reframe her thoughts and feelings as cues that she needs to be more open to differences.

Step 3. Don't Try to Fix Fixed Mind; Be Kind Instead

1. *Practice being open to what is happening in this moment. This includes being more open to feedback to determine if you should adopt it. Sandy considers her volunteers' point of view.* Sandy understands that someone could want a lunch break and still care. She continues to believe

that her friend could describe the energy bill better and decides to do trainings about it at the group's next meeting. She self-validates her frustration over one volunteer getting side-tracked by an attendee's inquiry into heat pumps and at the same time decides to practice acceptance and reframe it as a moment for connection, which might be more important for gaining membership than communicating the specific details of the bill being endorsed at that moment.

2. *Change Physiology. Closed mouth smile while breathing deeply; use eyebrow wags* [raising your eyebrows while smiling gently with a closed mouth]. By greeting her volunteers with nonverbals, Sandy found they were more agreeable and seemed to be in better spirits.
3. *Rather than resisting, fixing, or defending your Fixed Mind, allow it to be.* Trying to criticize yourself for being too critical doesn't work. Rather, respond to fixed mind with curiosity for what it can teach you about yourself and/or your relationships with others.
4. *Use a loving kindness practice.* As stated in the *Be Kind to Fixed Mind* RO DBT Handout: Repeat slowly to yourself, "May my Fixed Mind find peace, may my Fixed Mind be content, may my fixed mind be safe and secure."
5. *Forgive yourself for being in Fixed Mind.* This entails normalizing the experience of fixed mind. Think of everyone who has ever corrected something you did or said. They were in Fixed Mind, and it wasn't all bad.

Fatalistic Mind Case Example

Darren is a 25-year-old copy writer who is anxious about climate change and wants to do something to help. He isn't an environmental scientist, politician, or lawyer. However, he has felt stagnant in his career and sees an opportunity to reinvent himself by creating a working group of professionals to mobilize environmental activism. He is excited to move forward, but nobody shows up at his meetings despite his efforts to advertise the group. His colleagues seem disinterested and the grant proposal he writes for funding is rejected. Darren believes himself to be a failure, and thinks his efforts were a waste of time. He also resents his colleagues for not respecting his ideas.

Darren's therapist presents him with *Learning from Fatalistic Mind* (Lynch, 2018b, pp. 247–248).

Step 1. When Challenged, Observe Emotions, Action Urges, and Thoughts that May Be Linked to Fatalistic Mind

The handout lists several possibilities consistent with Fatalistic Mind, with Darren best fitting the following description: "Feeling unappreciated, invalidated, misunderstood, helpless like a small child, like a martyr, or like a victim AND/OR resentful, bitter, or cynical about change AND/OR numbed out or shut down" (p. 247).

Step 2. Acknowledge the Possibility of Being in Fatalistic Mind by Remembering that Fatalistic Mind Is the Opposite of Resisting or Fighting

Darren's therapist explains that Fatalistic Mind is the "escape artist" that thrives on feeling justified in walking away without acknowledging that you are avoiding facing challenges. His therapist also explains that Fatalistic Mind isn't necessarily bad. It simply means he might need to slow down or rest.

Step 3. Listen and Learn from Fatalistic Mind by Using the Following Skills

1. *Take the first step by acknowledging you don't want to.* Darren's therapist reminds him that he is choosing to behave fatalistically.
2. *Welcome despair, anxiety, and hopelessness as teachers, not enemies.* Darren's therapist encourages him to take ownership of his pain and also to not judge it. His therapist encourages him to sit with difficult feelings and reconceptualize them as teachers, not enemies.
3. *Let go of longing for the world to change or secretly hoping that the problem will go away.* Darren's therapist encourages him to focus on the potential for growth instead of blaming his colleagues.
4. *Remember, Fatalistic Mind does not mean that you are necessarily doing anything wrong.* Darren's therapist helps him consider that he might need a break and some rest, as well as time to reflect given that he feels stuck in his career and unsure how to channel his eco-distress constructively. It's OK to take a break because doing so will help him strategize and explore the best route moving forward.
5. *Turn your mind toward your predicament. Closed-mouth smile, eyebrow wag, and breathe deeply while thinking about the problem or feedback.* These simple behaviors can help change one's own perspective as well as positively influence others.
6. *Turn your mind to the possibility of change.* Darren was encouraged to seek guidance from a mentor, a colleague who has been able to obtain support for a beach clean-up project, and to be open to the mentor's feedback.
7. *Clarify the steps needed to solve the problem and then take the first step.* Darren is reminded that he doesn't have to have everything figured out now. All he needs to do is take the first step, which could simply entail talking with a mentor.

8. *Fatalistic mind thrives on secrecy.* Darren was encouraged to self-disclose to one safe person (ideally the mentor discussed above) about his urges to give up.
9. *Remember that rejecting help from others keeps you stuck in fatalistic mind.* This helped Darren remain open to his mentor's feedback and to seeking additional feedback from his colleagues.
10. *Forgive yourself for operating in fatalistic mind.* Forgiveness is a common theme across RO DBT skills. It is conceptualized as grieving the loss of an expectation, whether it be an expectation one had of others or of oneself. Forgiveness heals the forgiver.

WHEN ECO-DISTRESS INTERFERES WITH VALUED GOALS

In RO DBT, *valued goals* refer to those qualities that we want others to see us embody in interpersonal interactions. Examples include "To care for nature and the natural environment," "To forgive those who have harmed me;" "To be open-minded," "To not always assume I am right," "To not always assume I'm wrong," and "To acknowledge fallibility without falling apart" (Lynch, 2018b, pp. 226–228).

Sandy had mentioned that when she saw things weren't going as planned at the tabling event, she thought, "They just don't care enough." She also said she wanted to work toward the valued goal of others taking her seriously. An RO DBT therapist might then explore which social signals she showed at the tabling event. For example, instead of scheduling slots for each volunteer to be provided a lunch break, she rolled her eyes and sighed when one of them asked for it. Sighing and rolling eyes will not help reach her valued goal since people tend to respond to these behaviors by keeping their distance and not taking the other person seriously.

SELF-ENQUIRY FOR ECO-DISTRESS

Both RO DBT clients and therapists are expected to practice *self-enquiry*, and to strengthen their practice of a technique called *outing yourself*, in which you are transparent about a vulnerability such as taking responsibility for trying to avoid a difficult situation.

Here are some eco-related entries from my self-enquiry journal:

Sunday, December 31, 2023

I felt guilty about using plastic bottles last night at a neighbor's party and throwing them in the trash because my neighbors didn't have a place in their home to put recyclables. I could have brought them home to recycle. I didn't want to look weird,

> *but now I'm comparing myself to people who go without drinking water rather than drinking from plastic water bottles. Is it possible I'm experiencing unhelpful envy towards them?*

Notice that I end with a question, and particularly one that brings up uncomfortable truths, i.e. feeling jealous. Ending with a question is meant to prevent the client from unintentionally rationalizing, justifying, or self-soothing, which might prevent learning. For example, thinking that it's OK I didn't recycle because I was tired would prevent me from facing my eco-guilt, and acting defensively in response to the social comparison from water bottle abstainers, such as concluding that they are crazy, would prevent me from acknowledging my envy.

Here's one more entry:

> Sunday, October 15, 2023
>
> *Lots of eco guilt today. I let the water run while I was cooking. I am aware of thoughts of being selfish. What do I need to learn? Is it possible I am in Fatalistic mind?*

Self-enquiry is meant to be brief (1–2 min) and you are supposed to stop this exercise when you find yourself starting to come to conclusions or solutions rather than sitting with a question. Hence, in the above entry, I forced myself to stop writing when I made statements suggesting the conclusion of being in Fatalistic mind, or in the previous example, of experiencing envy.

One question often included in a self-enquiry entry is "What is it that I need to learn?" RO DBT encourages a continuous process of self-growth, which is a challenge for overcontrolled individuals who are driven to arrive at the final, correct answer as soon as possible.

RO DBT can help climate activists prevent burnout by balancing the need for productivity with the need to engage more in the process of their endeavors, which requires rest periodically so we can sustain the long haul.

This reminds me of the time my family and I hosted a garage sale to get rid of items that were collecting dust in our home. It was a sunny and warm Saturday and I shared with a customer, an older woman, how much I enjoyed the excuse to be outside rather than spending the day cleaning and cooking like I do most other weekends. The woman looked at me with empathy and said, "Oh honey. The dirt isn't going anywhere. It'll *still* be there tomorrow!" In other words, the endless cycle of clean, cook, repeat feels endless when we choose not to take breaks from it. In contrast, being outside, in nature, can be the very reset that gives us renewed energy and purpose in our endeavors.

RESEARCH ON RO DBT

RO DBT is a relatively new CBT intervention. However, a systematic review of RO studies that included two qualitative articles, one case study, five case series studies, four quasi-experimental studies, and two articles regarding one randomized control trial (Hatoum & Burton, 2024) revealed emerging evidence to support RO skills for both adolescent and adult populations for improving symptoms in disorders of overcontrol, such as anorexia nervosa, autism spectrum disorder (in individuals with low support needs), and treatment-resistant depression.

In addition, two single-group experimental studies included in the systematic review (Hatoum & Burton, 2024) showed decreases in global distress, obsessive-compulsive symptoms, anxiety, need for structure, and increases in functioning, wellbeing, perception of recovery, closeness, and ability to depend on others when needed (Cornwall et al., 2022; Egan et al., 2021).

To date, we lack research that specifically looks at the effectiveness of RO for eco-distress, nor is there research on how RO skills could be helpful within community settings in better engaging the public in environmentalism by reducing obstacles of perfectionism, maladaptive social signals, overcontrolled coping styles, and overcontrolled states of mind. My hope is that writing about the potential for widespread benefit could inspire more research.

By getting away from competition, perfectionism, and materialism, we can truly appreciate the value of strengthening genuine connections.

REFERENCES

Cornwall, P. L., Simpson, S., Gibbs, C., & Morfee, V. (2022). Evaluation of radically open dialectical behaviour therapy in an adult community mental health team: Effectiveness in people with autism spectrum disorders. *BJPsych Bulletin, 46*(5), 308308. https://doi.org/10.1192/bjb.2022.47

Curran, T. (2023). *The perfection trap: Embracing the power of good enough.* Scribner.

Curran, T., & Hill, A. P. (2019). Perfectionism is increasing over time: A meta-analysis of birth cohort differences from 1989 to 2016. *Psychological Bulletin, 145*(4), 410–429. https://doi.org/10.1037/bul0000138

Edmondson, A. C. (2023). *Right kind of wrong: The science of failing well.* Simon Element/Simon Acumen.

Egan, R., Long, E., McElvaney, J., & Booth, R. (2021). Group radical openness: A feasibility study. *Counselling and Psychotherapy Research, 22*, 913–924. https://doi.org/10.1002/capr.12480

Hatoum, A. H., & Burton, A. L. (2024). Applications and efficacy of radically open dialectical behavior therapy (RO DBT): A systematic review of the literature.

Journal of Clinical Psychology, 80(11), 2283–2302. https://doi.org/10.1002/jclp.23735

Lynch (2018a). *Radically open dialectical behavior therapy: Theory and practice for treating disorders of overcontrol.* Context Press.

Lynch (2018b). *The skills training manual for radically open dialectical behavior therapy: A clinician's guide for treating disorders of overcontrol.* Context Press.

Over Zero and The American Immigration Council. (2024). *The belonging barometer: The state of belonging in America.* Revised ed. https://www.projectoverzero.org/media-and-publications/belongingbarometer

Portland.gov. (2018–2025). *Saving water in your home.* https://www.portland.gov/water/water-efficiency-programs/save-water-home

Wray, B. (2022). *Generation dread: Finding purpose in an age of climate crisis.* Knopf Canada.

Chapter 7

Dialectical Behavior Therapy

Turning the Faucet Off on Eco-guilt, Anger, and Despair

The headline read: "Climate Activist Dies After Setting Himself on Fire at Supreme Court" (Cameron, 2022). While rare, there have been instances in which the threat of climate change, and the government's lack of action to protect us from it, provoked such intense hopelessness to motivate suicide. In this case, Wynn Bruce, a 50-year-old man from Colorado, set himself on fire in front of the Supreme Court on April 22, 2022, as part of an Earth Day protest. Whether it be guilt, rage, or profound despair, emotion regulation difficulties lie at the heart of suicidal behavior (e.g. Colmenero-Navarrete et al., 2021).

It's important to remember that negative emotions such as guilt are not always bad in the context of climate change. For example, recently I purchased a textbook and a pop-up asked me to consider the environment before purchasing it. This gave me pause, and I quickly shifted to buying an online version to avoid guilt and regret.

On the other hand, research has consistently highlighted trouble regulating negative emotions as a risk factor for suicidal behavior (e.g. Colmenlero-Navarrete et al., 2021). There are ways in which climate change threatens to exacerbate an already disturbing trend upward in suicides among teenagers and young adults (e.g. Verlenden et al., 2024). For example, one study completed in New South Wales found consistent associations of greater youth suicides on hotter days (Dey et al., 2024). Heat has a myriad of negative health consequences, many of which are relevant to mental health, including disruption of sleep, which in turn affects mood (e.g. communication with J. McKowen, August 8, 2025). Research (Wong et al., 2024) has also shown that people with severe mental illness, such as schizophrenia, are at greater risk of heat-related illness. Regardless of whether they were being prescribed psychiatric drugs.

DOI: 10.4324/9781003489665-8

Attempts at extreme solutions to the climate crisis are also bound to exacerbate emotional wellness. For example, while striving to use resources in moderation is great for the planet, operating in the extreme of restricting oneself of food and rest is self-defeating and is likely to lead to having less energy to be an environmental advocate.

In this chapter, I discuss dialectical behavior therapy (DBT), a type of cognitive behavioral therapy developed by psychologist Marsha Linehan (1993) that had the initial aim of targeting suicidal and non-suicidal self-injurious behaviors, such as cutting, burning, or hitting/punching/kicking oneself. In Linehan's conceptualization, self-injurious behaviors might serve as the client's attempt to regulate intense emotions that had been invalidated throughout their lives. The skills she developed for building a life worth living ultimately blossomed into a comprehensive skill set, with both individual and group treatment protocols, and it became the gold-standard treatment for borderline personality disorder (BPD).

BPD (Linehan, 1993) is a personality disorder characterized by profound emotion dysregulation, trouble with maintaining relationships, and self-injurious and suicidal behaviors. However, over the years, use of DBT has been broadened to treat other mental health conditions such as substance abuse and eating disorders. Beyond treating serious mental health disorders, skills from DBT, such as mindfulness, have been incorporated into garden-variety self-help tools that are widely available to the public.

A core concept of DBT is dialectics, a holistic perspective according to which opposing truths, called the thesis and antithesis, are interrelated, with neither alone being an absolute truth (Linehan, 1993). The therapist guides the client in understanding a synthesis between them, honoring their truth while also getting the client to shift toward change. One of my favorite examples of a dialectic, a mantra of mine for parenting, is "I am doing the best I can, and at the same time, I need to do better." Being able to synthesize dialectics is important in regulating emotions sufficiently to sustain relationships, all of which are critical ingredients to a stable sense of self-identity.

In discussing how applying DBT skills can help those experiencing the complex array of emotions related to the climate crisis, this chapter unpacks a dialectic. On the one hand, the biosocial theory, a conceptualization for emotional difficulties addressed by DBT, maps well onto the institutional invalidation of climate emotions, as evidenced by political inaction despite the gravity of the climate crisis (J. Homan, personal communication, November 4, 2023) and norms that scorn talk of the severe consequences of climate change. Given that those who are more sensitive to their surroundings are also more connected with nature and animals (Setti et al., 2022), and that people with an anxious temperament are more likely to be impassioned by environmental activism (Pickering & Dale, 2023), skills for managing intense emotions are in great need if we are to harness them into climate action.

On the other hand, advocating psychotherapy for climate change, including DBT runs the risk of pathologizing normal responses to a threat of such magnitude as climate change. I don't disagree with this perspective. However, I was concerned by some readers' comments to the *New York Times* piece reporting on the tragedy of Wynn Bruce's suicide (Cameron, 2022), framing it as a noble sacrifice for the climate cause. One reader compared him to Jesus Christ. These sentiments are unhelpful at best and toxic to anyone who is at risk of suicidal behavior.

I also don't agree with those who condemn individuals pushed to extremes like suicide, such as comments referring to him as being selfish and mentally ill. It is more helpful to respond with compassion and a mind toward prevention of future tragedies. The question we should be asking is, how do we help people who have lost hope regain it?

Another dialectic related to climate change is understanding that while we educate others on the destructive powers of consumerism and waste, we also continue to contribute to the mess. We are part of the very system we are begging the government to overhaul. I will be the first to admit that this system, one in which I am able to take vacations, and use modern appliances like dishwashers and dryers, is one that brings me much comfort and contentment. Every time I run a load of wash, I sit with the discomfort of my hypocrisy. Before you can tell me to join a commune of homesteaders, I'm already in defensive mode. But what about my children's education? What about my career, which requires a lot of time and therefore requires me to prioritize efficiency over sustainability when it comes to basic life necessities like preparing meals?

This dialectic of being part of the monster also comes up when I feel defeated in my sustainability efforts. For example, having to wash more sweaty clothing dampens the thrill of lessening my carbon footprint by biking to work. My mind is unable to calculate and compare astutely enough to know whether my efforts are worthwhile.

Statistician Paul Murtaugh and scientist Michael Schlax attempted to estimate a "carbon legacy" of lifetime emissions from a single person, taking into account that that person would be responsible for half of their children's and one-quarter of their grandchildren's emissions and so on and so forth (Hedburg, 2024). For example, while some might fret over the decision of whether or not to have children, others might feel comforted to know how much one could help the environment simply by having fewer children. As Trevor Hedburg, a philosophy professor at the University of Arizona who wrote an article about the research describes:

> If emissions stayed similar to 2005 levels for several generations, an American couple having one fewer child would save 9,441 metric tons of CO2-equivalent, according to their calculations. Driving a more fuel-efficient car, on the other hand – getting 10 more miles to the gallon – would save only 148 metric tons of CO2-equivalent. (para. 12)

DBT represents an ideal approach for eco-distress because a major aim of the treatment is to help the client self-validate their thoughts and emotions, as well as motivate their behavior in the service of their wise mind values. Wise mind (Linehan, 1993) represents the synthesis, or integration, of emotion and logic-driven thoughts and behaviors. According to Dr. Jesse Homan, a licensed professional counselor, someone with strong pro-environmental values, could engage in environmental sustainability without jeopardizing their well-being or relationships (Doherty et al., 2024).

Cultural norms prize materialism and frown upon "silly" concerns about turning the heat on higher than needed. Sacrificing comfort for sustainability is seen as masochistic and pathological and urging others to do the same may be met, at best, with an eye roll.

On the other hand, DBT allows us to understand the essence of emotion in driving human behavior, which means that fighting emotion with pure logic is insufficient. In this way, it has the potential to empower climate activists who are rightfully upset by government inaction.

We need not fear intense emotions. DBT serves to honor emotional intensity by teaching concrete skills for channeling these emotions in healthy ways without invalidating the individual. In doing so, we can send the message that emotional sensitivity is neither wrong nor pathological, and many of us need more of it to motivate us to stand up to the corporate greed and invalidating government officials who are exacerbating environmental devastation. People will say they care, but they need to show it more, and showing can be the emotionally sensitive person's superpower.

In its focus on skills for managing intense emotions and channeling them in a way that is empowering, DBT philosophically addresses the intense emotions that the existential threat of climate change understandably triggers, not just in improving the mental health of climate advocates but in encouraging all of us to allow ourselves to have intense feelings while at the same time managing them toward the end of promoting effective climate action.

DBT also emphasizes the link between caring for one's body and taming emotional intensity and reactivity. For example, PLEASE skills (Linehan, 2015) include: "Treat **P**hysica**L** illness, balance **E**ating, avoid mood-**A**ltering substances, balance **S**leep and get **E**xercise" (p. 247). By caring for ourselves, we are allowing ourselves to be more present in climate action while also helping our mental health.

USING DBT DEAR MAN SKILLS TO CHANNEL EMOTIONS INTO EFFECTIVE CLIMATE ACTION

DEAR MAN skills (Linehan, 2015) are part of the Interpersonal Skills Module of DBT. They are helpful in getting people to clearly communicate what they want/need from others, including when they need a boundary to be set. DEAR MAN, like other DBT skills, is presented as an acronym. Below I present each skill and apply it to scenarios related to the difficulties we have already discussed, such as trouble joining with other environmental advocates:

> **DEAR MAN**, a DBT skill in the Interpersonal Skills module goes as follows:
>
> **D**escribe; **E**xplain; **A**ssert; **R**einforce/Reward; **M**indful; **A**ppear confident; **N**egotiate.

First, you describe the situation as it is, sticking with the facts. For example:

> You are asking me to print out poster materials when I need to take my daughter to the doctor this afternoon. I will not be able to do this task, and I have offered to help in other ways. I am noticing that you are getting frustrated with me for not being able to complete the task you are asking of me.

Express your feelings and opinions without assuming that others know how you feel.

> I am feeling confused and hurt. I want to be part of this effort to advocate for environmentally friendly policies. At the same time, I want to be understood when I need to prioritize my family.

Assert yourself by asking for what you want or drawing clear boundaries.

> I can't do what you are asking of me.

R stands for Reinforce/reward. Praise the person in advance, emphasizing the positive effect their willingness to grant your request will have.

> I really appreciate you understanding why I can't print out materials before tomorrow's protest.

M stands for Mindful. Repeat your main points without getting off topic. Keep making your point while ignoring distractions and negative comments.

> As I said, I can come to the protest, and I will not be able to print out materials beforehand.

Appear confident in your verbal (clear, audible voice) and non-verbal (stand up straight with eye contact) communication.

Negotiate: Be willing to compromise. Focus on what will work rather than who won or lost or who was right or wrong.

> I can't print out the materials, and I can email our friend and ask her to help with this.

DEAR MAN skills come in handy in response to pressure from other advocates to take on more work. Time and again, the response (including my own at times) has been to quietly, or not so quietly, quit because it was "too much." DEAR MAN can help such individuals stay in the game and set appropriate limits. While I understand the urgency felt by activists and their frustration when others can't join a protest due to other commitments, we need to meet people where they're at regardless of their level of commitment. We are all a part of this crisis. There is no competition to win the Most Dedicated award, at least none that matter when it comes to surviving an existential crisis.

People in my personal life and clinical work have been pressured to call their school administration when they didn't feel equipped (with knowledge, political prowess, or mental space) to face public school officials. Some have been pressured on social media to watch graphic videos of people suffering as part of their protest. If they couldn't stomach it, they were told they didn't care enough. DEAR MAN can be used to clearly express limits. If your activist group digs their heals in, there are plenty of others who would be thrilled to have you join.

ACCEPTS SKILLS FOR COPING WITH INTENSE EMOTIONS

ACCEPTS is a set of skills from the Distress Tolerance module of DBT (Linehan, 2015). Distress Tolerance skills are meant to get clients through an emotional crisis. They are meant to be used temporarily and learned in conjunction with other skills that are more enduring across contexts. It could be helpful to someone feeling hopelessness and despair about climate change, rage over a politician's decision to endorse an oil drilling project, or news of a natural disaster.

A*ctivities – Do something* to get your mind off of what is upsetting you. Remind yourself that this is not only OK, but necessary in order to sustain pro-environmental efforts.

Watching or listening to comedy can help, even climate change related comedy. In one of my favorite comedic podcasts, *Conan O'Brien Needs a Friend,* Conan is clearly an advocate of environmental protection and playfully pokes fun at climate deniers in a way that allows us to sit with the earth's trouble in a more digestible, lighthearted way.

C*ontributing – Do something nice for somebody else.* You can cook for someone, possibly preventing them from eating out or needing to go to the store. You can tend to your or someone else's garden. In other words, you can do something for somebody that lessens both your and the other person's carbon footprint, or you can give to the earth directly, such as by planting a seed.

C*omparisons – Compare yourself to those less fortunate or compare yourself to a time when you were coping less well.*

Keep a journal of pro-environmental behaviors and your progress toward goals. Look back and focus on the *overall* progress you have made. Remember, occasional setbacks are to be expected.

E*motions – Create different emotions.* Spending time in nature inspires awe and helps with staying grounded in the present.

Over the past year of worrying about changes to agencies such as the Environmental Protection Agency (EPA), biking to work helped get my blood moving. Simply looking at my viewing queue on *Netflix* also gave me a sense of normalcy, and reading positive news, such as measures taken by states to protect environmental safeguards, have been helpful. I joined environmentally minded groups on Facebook local to my community, such as a sell and swap and a waste-reduction group.

P*ushing away – Push the painful situation out of your mind temporarily.* Mentally put emotional pain in a box and on the shelf.

Replace Your Thoughts

Remember, the emphasis is on getting through the moment, not actively coping, such as problem solving in the longer term. That's why these skills are meant to be used in conjunction with others.

Sensations – Intensify Other Sensations

Anyone who has spent more than a day with me knows that my go-to is caramel-coffee flavored sucking candies. When I'm stressed, I focus on the texture and sweetness of these candies.

MINDFULNESS OF CLIMATE EMOTIONS

Mindfulness skills allow clients to disengage from their emotions and thoughts, watching them with curiosity and without judgment (Linehan, 2025). When we think "I'm upset because I'm not doing enough to help the environment" we can imagine the words floating on a cloud observing them as separate from us as opposed to being fused with it as if it is part of who we are. Specifically, we can separate ourselves from the judgment of this thought by reframing it as "I am aware of a thought that I'm not doing enough to help the environment." We thus conceptualize thoughts as random events as opposed to automatically being facts.

Mindfulness skills in DBT are divided into "how" and "what" skills (Linehan, 2025). The how skills instruct us to practice mindfulness nonjudgmentally, such as not judging a feeling of anxiety or a thought about feeling hopeless; one-mindfully, or focusing on one thing at a time, and effectively, or to do what works. *How* skills guide people in how to practice *what* skills, which can be further divided into three categories: *Observing*, *Describing*, and *Participating* (Linehan, 2025). All can be applied to mindfulness practices in nature. I've found mindfulness to be helpful when used with intention during "forest bathing" (e.g. Farkic et al., 2021) or riding my bike on an autumn day. My clients have found solace simply by going to their backyards and observing leaves falling from trees.

Observing means noticing the properties of what we are experiencing through our senses without labeling them (Linehan, 2025). Examples include noticing a leaf gently dancing in the wind without putting words to it, focusing your attention on the sensation of your pet's fur as you stroke it, or noticing the heaviness in your soul without calling it sadness.

Describing skills involve labeling an experience but in a nonjudgmental matter, as mentioned above (Linehan, 2025). I describe a faucet running while I am loading a dishwasher without judging myself for waste. This doesn't mean I won't change my behavior to reduce waste. In contrast, noticing and describing the facts will allow me to adjust my behavior without unnecessary self-condemnation, which could distract me from being aware of other opportunities to be environmentally conscious. It will pave the way for creativity, such as when I realize I can reuse a takeout container to hold unmatched socks, which I could save for sock puppet crafts my kids could make on a rainy day.

Participating is becoming one with an experience, such as losing yourself in a dance or when singing a song (Linehan, 2025). You are fully engaged in what you are doing with reduced mental chatter. This is what I strive for when I'm spending time with loved ones or watching the sun rise.

Another important aspect of emotion regulation skills is being able to identify emotions. Mental health professionals have created various tools for naming climate-related emotions, such as the Climate Mental Health Network's *Climate Emotions Wheel* (Pihkala & Kamenetz, 2024).

DBT emphasizes allowing all emotions. While our actions in response to emotions have the potential to be destructive, emotions themselves are harmless and worthy of validation. They can be employed, furthermore, in the service of actions that are in line with wise mind values.

The Emotion Regulation module of DBT also includes skills for managing emotions in a way that allows you to be safe and effective in your endeavors to live in accordance with wise mind values for environmental sustainability. One such skill is known as Opposite Action (Linehan, 2025).

OPPOSITE ACTION FOR ECO-DISTRESS

In opposite action (Linehan, 2025), the client is guided in understanding whether the situation triggering the emotion would be helped by engaging in behavior you feel an urge to do in response to the emotion, such as running away from someone who is attacking you. If the behavior is adaptive, such as running away from an aggressive person, then it is likely best to engage in the behavior. However, if the behavior is unjustified and/or ineffective, opposite action guides the client to go opposite of what they have an urge to do.

For example, if you are anxious about an upcoming test, feeling anxious is justified but giving into the urge to avoid studying for it wouldn't be effective. This is when the time is right to practice *opposite action*. This means that even though your gut tells you to avoid the situation, you are going to lean into doing the exact opposite, approaching it.

CONCLUSION

DBT skills help us mold helpful reactions to situations based on contextual factors, which in turn have implications for how we respond to the climate crisis. For example, rather than wasting valuable energy exasperating ourselves with screaming at climate deniers, we can start to make persuasive appeals to people who are anxious about climate change but feel powerless over it. With climate deniers, we can listen to their side and try to understand it even if we don't agree with it. At the very least, we can hope that understanding their perspective will make it easier to prevent them from getting in the way.

Achieving this will require tapping into your own and others' values. For example, you might not convince someone to buy locally for environmental reasons, but attending a local farmer's market might be appealing for other reasons, such as healthier options, lower costs, socializing with others in the community, etc. Saving water is the desirable outcome of doing less laundry, and doing less laundry means more time for leisure, which could be a selling point even for those who aren't as concerned about wasting water.

The importance of exploring your own values and learning to better tolerate people who have different sets of values and priorities is a major emphasis in the cognitive behavioral intervention we explore next, Acceptance and Commitment Therapy (ACT).

REFERENCES

Cameron, C. (2022, April 24). Climate activist dies after setting himself on fire at Supreme Court. *The New York Times*. https://www.nytimes.com/2022/04/24/us/politics/climate-activist-self-immolation-supreme-court.html

Colmenero-Navarrete, L., García-Sancho, E., & Salguero, J. M. (2021). Relationship between emotion regulation and suicide ideation and attempt in adults and adolescents: A systematic review. *Archives of Suicide Research, 26*(4), 1702–1735. https://doi.org/10.1080/13811118.2021.1999872

Dey, C., Wu, J., Uesi, J., Sara, G., Dudley, M., Knight, K., Scott, J. G., Jay, O., Bowden, M., & Perkes, I. E. (2024). Youth suicidality risk relative to ambient temperature and heatwaves across climate zones: A time series analysis of emergency department presentations in New South Wales, Australia. *Australian & New Zealand Journal of Psychiatry, 29*(1), 18–28. https://doi.org/10.1177/00048674241290449

Doherty, T., Artman, S., Homan, J., Keluskar, J., & White, K. E. (2024). Environmental identity-based therapies for climate distress: Applying cognitive behavioural approaches. *The Cognitive Behaviour Therapist, 17*, e29. https://doi.org/10.1017/S1754470X24000278

Farkic, J., Isailovic, G., & Tyalor, S. (2021). Forest bathing as a mindful tourism practice. *Annals of Tourism Research Empirical Insights, 2*(2), 1–9. https://doi.org/10.1016/j.annale.2021.100028

Hedburg, T. (2024). Children are expensive—not just for parents, but the environment—So how many is too many? *The Conversation*. https://theconversation.com/children-are-expensive-not-just-for-parents-but-the-environment-so-how-many-is-too-many-176662

Linehan, M. (1993). *Cognitive-behavioral treatment of borderline personality disorder*. Guilford Press.

Linehan, M. (2015). *DBT skills training handouts and worksheets* (2nd ed., revised). Guilford Press.

Linehan, M. (2025). *DBT skills training handouts and worksheets* (2nd ed., revised). Guilford Press.

Pickering, G. J., & Dale, G. (2023). Trait anxiety predicts pro-environmental values and climate change action. *Personality and Individual Differences, 205*, 112101. https://doi.org/10.1016/j.paid.2023.112101

Pihkala, P., & Kamenetz, A. (2024). A guide to climate emotions. *Climate Mental Health Network*. https://www.climatementalhealth.net/wheel

Setti, A., Lionetti, F., Kagari, R. L., Motherway, L., & Pluess, M. (2022). The temperament trait of environmental sensitivity is associated with connectedness to nature and affinity to animals. *Heliyon, 8*, 309861. https://doi.org/10.1016/j.heliyon.2022.e09861

Verlenden, J. V., Fodeman, A., Wilkins, N., Everett Jones, S., Moore, S., Cornett, M. S., Sims, V., Saelee, R., & Brener, N. D. (2024). Mental health and suicide risk among high school student and protective factors—Youth risk behavior survey, United States, 2023. *MMWR Supplement, 73*(Suppl4), 79–86. http://dx.doi.org/10.15585/mmwr.su7304a9

Wong, A. Y. S., Iwagami, M., Taniguchi, Y., Kawamura, C., Suzuki, A., Douglas, I. J., Bhaskaran, K., Sugiyama, T., Kuroda, N., Nitsch, D., & Tamiya, N. (2024). The role of psychotropics on the associations between extreme temperature and heat-related outcomes among people with mental health conditions: Population-based study. *Psychological Medicine, 54*(16), 4658–4664. https://doi.org/10.1017/S0033291724002824

Chapter 8

Acceptance and Commitment Therapy

Getting in Tune with Values

THE "I'LL DO IT WHEN …" PARADOX: BASIC PRINCIPLES OF ACT AND HOW THEY ARE RELEVANT TO ECO-DISTRESS

Acceptance and commitment therapy (ACT) is a third-wave cognitive behavioral therapy that focuses on how we manage our responses to uncomfortable thoughts and emotions (Twohig et al., 2021). ACT helps people relate better to their emotions and thoughts by perceiving them as informative signals rather than threats. The target of treatment is psychological inflexibility, which manifests in the person being too attached to their internal experiences, such as erroneously assuming that because one feels ineffective, they must actually be so. The overarching goal is to allow the client to more effectively approach distressing experiences so that they can live a more meaningful and valued life (Sabucedo, 2021).

In aiming to help people commit to actions that are in line with their values, ACT techniques can be helpful to individuals who want to do something to help humanity but feel distressed when they come up against obstacles to their attempts to do so. Consistent with this observation, Generation Z employees prioritize meaningful work (Brower, 2024), but even this perspective can be tainted with inflexibility if they are let down about the inevitable grind that most jobs at least occasionally entail, like completing time sheets and other administrative tasks. They might narrow their options to jobs, such as for non-profit organizations, that have a humanitarian mission statement or find that launching projects that are outside of everyday business is daunting given the additional time and energy they take along with an uncertain outcome.

DOI: 10.4324/9781003489665-9

An example of a sign of inflexibility is the belief that one must first accrue wealth, time, and or more knowledge before they can lean into creative and/or humanitarian endeavors. The mentality of "First I must have X before I can consider Y" creates rules and contingencies for taking prosocial action, including climate action.

As mentioned earlier in this book, individuals who are more risk-averse, as is often seen in anxious people, tend to hesitate to take action if they aren't sure how successful they will be or how they will achieve the desired outcome. This is another example of placing a contingency on acting.

ACT can help people identify components of psychological inflexibility (Twohig et al., 2021). For example, the concept of *cognitive fusion*, which ACT seeks to reduce, means responding to a thought or feeling as if it reflects absolute truth. "I can't pursue this work unless I know for sure that it will have a significant impact" is thus followed like a law, even though its power is completely in the mind of the individual. This isn't a delusion, and in fact often these kinds of thoughts reflect reality to an extent. Rather, we are focusing on their judgmental tone. For example, if someone thinks "I'm unintelligent," the thought itself doesn't mean anything about their actual intelligence. Even though the individual knows that it is their interpretation, sometimes the feelings and actions that follow in response to the thought are in line with the assumption that the thought is true, such as avoiding challenging tasks they assume they are incompetent at.

ACT helps people take a step back from their thoughts through a skill known as *cognitive defusion*, which could be as simple as reframing the thought "I'm unintelligent" with "I am noticing a thought that I'm unintelligent," thus creating some distance from the mental event as opposed to joining with it (Twohig et al., 2021).

What thoughts and feelings get in the way of climate action? Clients with eco-distress might be engaged in environmental activism, but not see their efforts bring about sufficient change. They might get hooked (the ACT term for perseverating) on thoughts of "What's the point?!" or "Nobody believes/will listen to me." They might also get frustrated with others in their group who are not as committed or unable to give as much time and energy to their cause.

People who want to do something about climate change might also face obstacles such as lacking time, believing themselves unknowledgeable or incapable of activism, or worrying that activism will get them in trouble with their jobs or the law. This is distressing, but because such obstacles are deemed outside their control, many people might not initially bring them to their therapist's attention. However, such conflicts between personal values and lack of committed action come up in more subtle ways. For example, they might be dealing with a partner that doesn't want them to spend their time doing advocacy work or with a job that is too demanding to leave room for volunteering.

In other cases, symptoms of mental health problems conflict with the client's environmental values, leading to greater distress. For example, someone with a diagnosis of obsessive-compulsive disorder might have a cleaning compulsion that conflicts with environmental values for using fewer resources. A client with Generalized Anxiety Disorder, characterized by excessive worries, might realize that their concerns about grades, work performance, and relationships are less meaningful in light of the threat of climate change, yet they still struggle to stop worrying about everyday concerns. Such conflicts could lead to self-inflicted shame. The goal of therapy then becomes twofold: To help the client nurture self-compassion and to guide them toward living in line with their values, including environmental ones.

Helping a client fulfill psychological needs such as belonging, meaning, and purpose might entail gently encouraging them to see how their symptoms are interfering with the potential to be part of a cause, whether this be individual pro-environmental behavior or collective action with climate-focused groups.

This is far from simple. An individual will inevitably and perpetually experience a values conflict with which they repeatedly make peace. This is because we have been born into unworkable systems that render us dependent on fossil fuels.

The following section focuses on the importance of addressing conflicting values, particularly when such conflicts affect one's ability to engage in collective action, such as joining with other activists or writing a letter to a representative.

CONFLICTING VALUES

Sometimes, thinking about the carbon footprint of the food I consume seems to easily lead to obsessing over it with a barrage of distressing thoughts, such as: With the vicissitudes of health advice, like whether we should eat organic, do I really need more to think about when it comes to what I eat? I already prefer vegetarian foods, but I eat a lot of dairy, which does not get a great environmental score (e.g. Griffiths, 2024). However, I have trouble digesting several plant protein foods.

While striving to reduce one's contribution to climate change is admirable and necessary, flexible thinking is also what is needed given the overwhelm that can come with becoming too fixated on reducing one's carbon footprint. For example, in *Inconspicuous Consumption: The Environmental Impact You Don't Know You Have,* author Tatiana Schlossberg (2019) speaks to the complexities of understanding the full effect our food production has on the planet. She writes, "Even organic meat might eat organically grown feed from somewhere else, or the manure to grow that feed came from far away" (p. 110). This highlights a concept called supply chain, or the network of raw materials, components, and goods that go into a single product and the journey they take before being assembled into that product (McKinsey & Company, 2022 August 17).

Schlossberg (2019) points out that ultimately the problem lies in food and transportation systems that are beyond any individual's control. At the same time, these systems are built to accommodate modern trends in consumption that were nonexistent in the past, such as year-round consumption of produce regardless of whether it out of season.

It's easy to see how, at least for some people, perhaps for many, eco-conscious eating can at best lead to frustration and at worst, down a risky path of restrictive, disordered eating. This is one of many examples illustrating how living in line with one's values is easier said than done. This is the reason why, in line with flexible thinking, I have found a balance by focusing on reducing food waste and doing so in creative ways such as finding new recipes for leftovers.

One problem I've seen in my clinical practice is that, in an effort to live in accordance with values, the client rigidly adheres to them. Ironically, this rigidity gets in the way of fulfilling their potential to express that value, such as by leading to conflicts with other values they prioritize. ACT offers tools for this issue by directly improving psychological flexibility, including flexibility in how we approach our values. Encouraging the client to think holistically and supporting them in effectively placing thoughts, feelings, and perceptions in context are two strategies that I have found helpful both personally and professionally.

Awareness can feel dreadful, such as when I learned not only about the environmental impact of chocolate but also about the humanitarian problem associated with how it is produced. What has been comfort food for me is a badge of privilege of which much of the world is deprived. I watched a disturbing documentary about the cocoa industry, an episode of the Netflix Series *Rotten* called *Bitter Chocolate* (Harper et al., 2019). The episode documents the history of exploitation tied to the cocoa industry, from children trafficked to work on cocoa farms in West Africa, to ongoing illegal child labor driving the work on these farms to this day.

Cocoa farmers and their children live on less than $1 per day, while the global chocolate market was valued at $129.6 billion in 2025 (Global Market Insights, 2026; Green America, 2019). African cocoa farmers and their families are trapped in poverty and see growing more cocoa as the only way out. Yet they have never tasted chocolate. The cocoa industry is also responsible for devastating deforestation. The Ivory coast has lost about 85% of its trees to make room for cocoa plants.

Luckily, Green America (greenamerica.org) has a "Child labor chocolate score card" on its website that can be used as a guide for selecting more ethical brands. According to the scorecard, chocolate retailers like *Lindt* and *Godiva*, the kinds you could easily pick up at a grocery store when you are grabbing a holiday gift for your child's teacher at the last second, are less than stellar performers on this metric. In fact, *Godiva*, a staple of many American shopping malls, gets an F. In a rushed moment, I nearly made it to the checkout aisle with a box of *Godiva* before I pivoted and returned it to the shelf as thoughts of child labor pounded through my mind. At that moment, it didn't matter to me if I looked strange or if people thought it was silly since one person cannot stop a corrupt system. But still, I wanted to do the right thing.

Some readers might poo-poo all of this as a first-world problem. It is, but it is also important to remember that the fear of *feeling* deprived is one that is shared by both environmental advocates and climate deniers. It is one that cannot be ignored.

Tony's Chocolonely was an ethical alternative to the brands I had been purchasing (Haynes, 2025). Tony's history is founded on the premise of creating an ethical chocolate brand that recognizes global inequality. The word "Lonely" is embedded in the brand's name because Tony felt lonely in his fight against the unethical aspects of the cocoa industry. The pieces of their bars look like a puzzle with unequal sizes and contours. This is done intentionally to represent wealth inequality. In addition, Tony's is widely available. While sustainable chocolate does not tend to be sustainably priced, Tony's was more feasible than other brands.

I thought I was all set in soul satisfaction, living in accordance with my values for humanitarianism and environmental sustainability while still enjoying chocolate, that is, until I learned about high concentrations of lead in dark chocolate (e.g. Callahan, 2023), which I had been eating by the fistful every day. A little more digging into the literature revealed the worst to be true. While brands like *Tony's Chocolonely*, *Theo*, and *Alter-Ego* were high achievers on Green America's ethical scorecard, these brands essentially scored at the bottom when it came to their lead levels.

The scientific community's recommendations for chocolate consumption are somewhat inconsistent (e.g. Callahan, 2023). Some experts reassure the public that they can continue to enjoy chocolate in moderate amounts and that people would have to consume inordinate amounts to experience any deleterious health outcomes. Others warn that even moderate amounts of lead can build up over time. *Consumer Reports*, which published an investigation (Loria, 2023) that was not peer reviewed (Callahan, 2023), list categories of chocolate brands high in lead and cadmium. The safer choices were either difficult to find, too expensive, or not included in Green America's ethical scorecard.

Needless to say, all of the thought put into what appears to be a far simpler decision for other people was taxing my limited time and energy. Ultimately, I purchased carob chips in bulk, as they are healthier and more sustainable (e.g. Gioxari et al., 2022). They are also pricey, so I mix this up with store-bought chocolate.

Resolving a values conflict such as this one isn't straightforward. We can take a look at ACT principles and interventions to see how they could apply to distress fueled by such conflicts.

For one, being flexible means looking at situations holistically, such as valuing my efforts to keep sustainability in mind when I purchase food, while knowing that staying healthy by prioritizing eating enough (calories, variety, etc.) and feeling emotionally satisfied when I eat means I won't be perfectly sustainable.

According to Kelsey Hudson, a clinical psychologist who practices ACT and has worked with and written extensively about helping youth with eco-distress, it is important to consider the role of each emotion and to ask yourself how emotions signal the way you are showing up to the situation (personal communication, January 10, 2023). For example, shame and guilt signal the urge to avoid or hide from others' attention and possible exposure of norm-breaking behavior. When it comes to those with eco-distress, this could show up in conflicting ways. On the one hand, the shame of going against norms consistent with sustainability, and, on the other, shame and embarrassment for standing out as vulnerable among those who don't prioritize living with a smaller carbon footprint and might not get it.

Anxiety, a common emotion experienced by those with eco-distress, serves the adaptive function of motivating action to prevent an adverse outcome. Sadness and hopelessness might indicate the need to find a supportive community and, in some cases, the need to slow down and tend to self-care to restore mental, emotional, and physical well-being.

Thomas Doherty, author of *Surviving Climate Anxiety: A Guide to Coping, Healing and Thriving* (2025), describes the importance of acknowledging negative emotions about climate change and understanding where they come from, while also welcoming additional, positive feelings, such as awe and wonder from observing nature. Doherty also emphasizes the need for prioritizing self-care as a coping mechanism for climate despair. During the summer of 2020, he helped his clients cope with the COVID-19 pandemic, and environmental threats in his area including heat waves and wildfires, by developing a technique called "The Upside Down Pyramid Exercise."

The Upside-Down Pyramid (Doherty, 2025) depicts essential needs like diet and rest at the bottom to indicate that these are the things we need to attend to more frequently, while activities like engaging in activism are on the top, also important but to be attended to after basic needs have been met. It is a visual way of assessing whether you are meeting your basic self-care needs and emphasizing that you need to tend to these in order to effectively be part of larger causes. The pyramid models Abraham Maslow's hierarchy of needs, according to which people generally (with some exceptions) address basic physiological needs, such as for sustenance and shelter, and basic emotional needs, such as for belonging, before they can pursue *self-actualization*, or fulfilling one's potential (Nevid, 2022).

Similar to DBT and RODBT, ACT also encourages enhancing awareness of bodily sensations, thoughts, and emotions by observing these phenomena and thinking of them as separate from yourself as a person (Twohig et al., 2021). In the ACT intervention known as NOTE, the acronym stands for **N**otice what is happening, including thoughts and feelings and your response to them, like whether you are avoiding them; **O**bserve your thoughts and feelings for what they are, with an emphasis on observing them with curiosity and without judgment; **T**urn toward your values, asking yourself how you would want to be remembered in terms of how you acted at this moment; and **E**ngage in what matters, including choosing to act in a way that matters most to you as per your values.

Applying this to my chocolate situation, I notice thoughts that I am causing others harm by purchasing chocolate with questionable ethical standards, and at the same time, my family enjoys chocolate, and I would rather spend more time with them than making hunting down the most sustainable chocolate my full-time job. In this case, understanding the small role you play in harm works to your benefit, and showing enough psychological flexibility to shift to a different value at that moment will help sustain you for the bigger causes you want to be a part of. In my case, I could encourage others to watch the documentary or request that more ethical brands be made available at local stores.

Another ACT intervention known as *Dropping the Anchor* operates on three principles captured by the acronym ACE, which stands for: **A**cknowledge your thoughts and feelings; **C**ome back to your body; and **E**ngage in what you are doing (Harris, 2019). In encouraging clients to focus awareness on internal events as phenomena rather than fact and to shift attention toward action, the ACT therapist is allowing them to get "unhooked" from the unhelpful thoughts and emotions that are leaving them feeling stuck.

THOUGHTS AND THOUGHT PATTERNS RELATED TO CLIMATE CHANGE

As described above, *Defusion* is an ACT concept that entails seeing thoughts and other internal experiences for what they are, signals one uses as information rather than something to get rid of or to fix (Twohig et al., 2021). Relatedly, the idea of a flexible sense of self, or *Self-As-Context*, refers to a perspective in which the self is seen as a container of inner experience while being distinct from it.

Defusion can be used alongside another ACT concept called *creative hopelessness*, according to which there is hopelessness in controlling inner experiences such as thoughts and emotions and at the same time an opportunity for creativity in being open to alternative ways of approaching problems.

In ACT, the therapist helps the client figure out what internal experience they're struggling with, what the client has done to try to control these internal thoughts or feelings, and then explore how effective these responses have been (Twohig et al., 2021). When the client sees that their attempts at control have caused more problems than the emotions and thoughts themselves, acceptance is presented as an alternative. Experiential exercises are used to help the client apply acceptance skills, such as writing a thought on a piece of paper and having them see the difference between holding the paper up to their faces, which represents how being stuck on the thought is blocking them from effective action, versus holding the paper on their lap which represents not fighting the thought. Throwing out the paper might seem like the natural solution, but trying to get rid of unhelpful thoughts often makes them even stronger.

ACT maintains that we cannot change other people's behavior or beliefs just like we cannot change who boards a public bus at each stop (Hayes, 2020). While we don't have to like the other person's beliefs, we can accept that their stance is part of reality at present. This can be helpful to people who are having trouble tolerating differences in those who haven't gotten on board with climate activism, or for those who are having trouble getting along with other activists who might assess the importance of one pro-environmental action over another differently.

ACT also emphasizes keeping your mind on the present, or what is happening in the world around you (Twohig et al. 2021). At the same time, the magnitude of the threat posed by climate change is real and it is of utmost importance that the clinician not unintentionally invalidate the client's concerns. This can be accomplished by first validating the client's emotions, then engaging them in cognitive defusion, and finally having the client observe experiences in the present moment with a stance of acceptance as opposed to engaging in experiential avoidance, which can worsen distress. Here's an example of a hypothetical interchange between a client and therapist in which these skills are applied:

Client: We have very little time left to do something about climate change before we have miserable lives. And nobody is doing anything about it!

Therapist [Validation]: You are describing a difficult situation. What kind of emotions come up for you when you think, "Nobody is doing anything about it?"

Client: I feel devastated and hopeless that nobody cares.

Therapist: The idea of nobody caring understandably brings up feelings of loss. What do you think these feelings of hopelessness and devastation might be telling you?

Client: They are telling me that maybe I should give up.

Therapist [Acceptance]: Can we just make space for this feeling of loss to be here? [Present-Mind]: What are you noticing right now as these emotions come up?

Client: I feel a heaviness and tightness in my gut. I want to cry and scream.

Therapist: Good job describing how they are coming up. And you took it a step further by telling me what actions these emotions and sensations compel you to do: cry and scream.

Client: Thanks. I still feel bad though.

Therapist [validation followed by guidance in cognitive defusion]: It's understandable to think you should give up when you feel hopeless, and for these thoughts to lead you to want to cry and scream. What if we were to try thinking of things you can do, even small things, that could help you build hope again?

According to Kelsey Hudson, it can be helpful to educate clients about opportunities for climate action that are happening locally (personal communication, December 3, 2024). Hudson shared that surrounding yourself, both physically and mentally, with people who also care about the environment, can be a valuable coping strategy. This could be accomplished by following networks that post positive stories about effective climate-related actions around the world.

ENVIRONMENTAL VALUES

Clinical psychologist Thomas Doherty integrates ACT with the idea of expressing environmental identity in a way that aligns with the client's eco-values (Doherty et al., 2024). To facilitate exploring environmental identity, Doherty asks his clients to draw a simple timeline from birth onward and note events that the client believes to be significant in their relationship with nature. In a different exercise, Doherty asks his clients to draw a family tree and to explore the nature connections of our relatives, such as predecessors who lived on farms or enjoyed bird watching.

Whether you are drawn to environmentalism because you are worried about the future of our species, concerned about animals, or invested in conserving beautiful forests and greenery, you have reason to be part of the environmentalists' cause. More specifically, you are already, inevitably a part of it, and the choice lies more so in whether you engage actively in a solution for preserving it.

Eco-values have been studied and divided into three categories (Schultz, 2000): Biocentric values entail concerns related to animals, plants, marine life, and birds. Egoistic values entail concerns related to self, one's future, lifestyle, and health. Altruistic values entail concerns for all people, children, and people in the community. The idea behind eco-values is that one's care for the environment is tied to sense of self and to the degree to which you believe yourself to be interconnected with others and nature. These categories were created as a result of a factor analysis based on participants' rank ordering the consequences of environmental harm that most concerned them. They can be applied to ACT's values assessment exercise, in which clients are guided in understanding how they wish to express their values. For example, someone with altruistic eco-values might want to talk to their school board about the health hazards of the combustion from gasoline-fueled buses, while someone with biocentric values might want to advocate for protecting endangered species.

In the *Bulls Eye* exercise, ACT therapists help clients understand the extent to which they are living in accordance with their values, pointing out the distance between how they want to live, which is represented by the bulls eye of a hypothetical darts board, and how they are living at present, which typically lands somewhere outside of the bulls eye (Twohig et al., 2021). The therapist guides the client to understand how this discrepancy is contributing to their distress and collaboratively works to identify committed actions the client can take to move closer to their values.

Furthermore, experiential exercises help the client understand factors that are creating the distance between how they wish to express their values and how they are expressing them through their current behaviors. These exercises, in turn, can help forge new environmental identities.

For example, I value being active, but I used to avoid going out in cold weather. Imagining how I would navigate the discomfort of the cold if my fear of it wasn't an issue, and without getting rid of feeling cold completely (which is impossible), motivated me to try snow shoeing. Enjoying this activity, in turn, allowed me to re-connect with the beauty of the natural world during the winter months and strengthened my gratitude for nature. Gratitude for nature has been shown to be associated with more willingness to engage in pro-environmental behavior (Tam, 2022).

CONCLUSION

It is understandable to feel saddened, frightened, and enraged by the lack of attention that our larger social institutions have given to the existential threat of climate change, and by the efforts of businesses to place responsibility on the individual, rather than themselves, for fixing climate change.

At the same time, taking responsibility for how we respond to this crisis, without getting hooked on nihilistic thoughts, will build resilience and help the planet.

Whether you are the climate activist who feels too burnt out to organize more protests, the working parent who wants to avoid food waste but is too overwhelmed by daily responsibilities, or the college student who is worried about climate change but whose perfectionism keeps you glued to obsessive workaholism, ACT can provide profound ways to help people get unstuck from internal chaos so they can engage more creatively in restoring order to the external mayhem of our current climate crisis.

REFERENCES

Brower, T. (2024, April 9). The Gen Z effect – and how the youngest employees are shaping the future. *Forbes*. https://www.forbes.com/sites/tracybrower/2024/04/09/the-gen-z-effect/

Callahan, A. (2023, February 9). Do I need to avoid dark chocolate now? *The New York Times*. https://www.nytimes.com/2023/02/09/well/eat/dark-chocolate-metal-lead.html#:~:text=study%20found%20that%20dark%20chocolate,%2C%20had%20much%20lower%20concentrations

Doherty, T. (2025). *Surviving climate anxiety: A guide to coping, healing, and thriving*. Little, Brown, Spark.

Doherty, T., Artman, S., Homan, J., Keluskar, J., & White, K. E. (2024). Environmental identity-based therapies for climate distress: Applying cognitive behavioural approaches. *The Cognitive Behaviour Therapist*, *17*, e29. https://doi.org/10.1017/S1754470X24000278

Gioxari, A., Amerikanou, C., Nestoridi, I., Gourgari, E., Pratsinis, H., Kalogeropoulos, N., Andrikopoulos, N. K., & Kaliora, A. C. (2022). Carob: A sustainable opportunity for metabolic health. *Foods*, *11*(14), 2154. https://doi.org/10.3390/foods11142154

Global Market Insights. (2026, April). *Chocolate market size & share 2026-2035*. https://www.gminsights.com/industry-analysis/chocolate-market

Green America. (2019, October 16). Child labor in your chocolate? Check our chocolate Scorecard. https://www.greenamerica.org/end-child-labor-cocoa/chocolate-scorecard

Griffiths, C. (2024, July 26). Dairy industry aims for GHG neutrality. Food & Water Stories. *The Nature Conservancy*. https://www.nature.org/en-us/what-we-do/our-priorities/provide-food-and-water-sustainably/food-and-water-stories/roadmap-collaborative-dairy-sustainability/#:~:text=Growing%20crops%20for%20cow%20feed,is%20disabled%20in%20your%20browser

Harper, A., Schillinger, T., Cechin-De La Rosa, C. (Writers), & Harper, A. (Director). (2019, October 4). Bitter chocolate. Season 2, Episode 5 [Netflix series episode]. In T. Schillinger (Ed.), *Rotten*. Netflix; Zero Point Zero.

Harris, R. (2019). *Dropping anchor: A script*. https://www.actmindfully.com.au/upimages/Dropping_anchor_script.pdf

Hayes, S. C. (2020). *Acceptance and commitment therapy: Principles of becoming more flexible, effective, and fulfilled*. Sounds True.

Haynes, S. (2025, April 6). Tony's Chocolonely: Leading the way in fair trade and social impact. *The Cincinnati Herald*. https://thecincinnatiherald.com/2025/04/06/tonys-chocolonely-fairtrade/#:~:text=%E2%80%9COur%20mission%20is%20to%20end,are%20created%20by%20Sterling%20Haynes

Loria, K. (2022, December 15). Lead and cadmium could be in your dark chocolate. *Consumer Reports*. https://www.consumerreports.org/health/food-safety/lead-and-cadmium-in-dark-chocolate-a8480295550/

McKinsey & Company. (2022, August 17). What is supply chain? https://www.mckinsey.com/featured-insights/mckinsey-explainers/what-is-supply-chain

Nevid, J. (2022). *Essentials of psychology: Concepts and applications* (6th ed.). Cengage.

Sabucedo, P. (2021). Acceptance and commitment therapy (ACT) and humanistic psychotherapy: An integrative approximation. *British Journal of Guidance & Counseling, 49*(3), 347–361.

Schlossberg, T. (2019). *Inconspicuous consumption: The environmental impact you don't know you have*. Balance.

Schultz, P. W. (2000). Empathizing with nature: The effects of perspective taking on concern for environmental issues. *Journal of Social Issues, 56*(3), 391–406. https://doi.org/10.1111/0022-4537.00174

Tam, K.-P. (2022). Gratitude to nature: Presenting a theory of its conceptualization, measurement, and effects on pro-environmental behavior. *Journal of Environmental Psychology, 79*, 101754. https://doi.org/10.1016/j.jenvp.2021.101754

Twohig, M. P., Levin, M. E., & Ong, C. W. (2021). *ACT in steps: A transdiagnostic manual for learning acceptance and commitment therapy*. Oxford UP.

Chapter 9

Standard Cognitive Behavioral Therapy

Revising Beliefs about Ourselves, the World, and the Future

HOW ARE YOU COPING?

"How are you coping?" was the way a former colleague started Zoom meetings during the Covid-19 lockdown. The conventional "How's it going?" seemed inadequate during those early weeks of the global pandemic, as my children, aged three- and five-years-old and quarantined from school at the time, interrupted every sentence with the roar of a T-Rex or a request for lunch at 10 am.

Our collective experience during the pandemic wasn't normal, hence the term the "new normal." Coping, or the process of using thoughts and actions to manage stressful or difficult situations (Merriam-Webs, n.d.), became the modus operandi.

Coping skills are also at the core of cognitive behavioral therapy (CBT), an umbrella term to describe interventions that have proven to be effective for treating a range of psychological issues, including anxiety, mood, substance-use, eating, and obsessive-compulsive and related disorders (Hofmann et al., 2012). There is also a considerable research base for using CBT to help individuals with somatic conditions such as insomnia and pain-related disorders.

CBT emerged in the 1960s as a reaction to the behaviorist approach that focused on measuring and modifying observable behavior (Dozois et al., 2019; Nevid, 2022). Early CBT theorists emphasized the critical role that thoughts also play in driving complex human problems. Furthermore, the CBT model (Beck Institute for Cognitive Behavior Therapy, 2025) illustrates the interconnectedness of internal events or thoughts, emotions (including physiological arousal of the autonomic nervous system) and observable

DOI: 10.4324/9781003489665-10

behavior. We see the world through the lens of our interpretations and feel and act accordingly. This determines how we interact with the world as well as how we cope with the pressures of it. Interpreting my children's interruption of meetings as unpleasant but manageable made the difference in whether I attended work meetings at all.

American Psychiatrist and professor Aaron Beck is globally recognized as the founder of CBT (Beck Institute, 2025). He described the role of *automatic thoughts* in how people respond to situations and how such thoughts determine behavioral responses. The hypothetical example I provide to clients when explaining CBT is one in which two individuals have a car accident. Both walk away without any injuries, but their cars are totaled. Person A has the thought that this is the worst day of their life given the inconvenience and expense of the crash. Person B, in contrast, thinks it is the best day of their life as they were lucky enough to walk away without a scratch.

According to the cognitive model, even though both individuals experienced the same circumstance, the differing nature of their automatic thoughts will likely lead to different emotional and behavioral responses. Person A is more likely to spend the rest of the day sulking, avoiding social interactions, and engaging in maladaptive behaviors such as drinking alcohol.

Ironically, this hypothetical blueprint resembles an event from my own life when I was a young driver and made a left turn into a busy intersection and crashed into a car I hadn't seen coming in the opposite direction. The other driver seemed frazzled but more forgiving than I had feared, while I started crying out of remorse. My upset, while understandable, was exacerbated by *cognitive distortions*, first described by Beck in a 1963 paper, specifically *catastrophizing* the situation as I thought "I'm a terrible person!" (Whalley, 2019). I also inappropriately placed blame on others for my being in a rush that day, an example of *personalization*, which entails placing disproportionate blame on either yourself or others. I later came to learn that the other driver had survived a life-threatening disease earlier in their life and that helped them put the incident in perspective.

In the context of eco-distress, ruminating over leaving the lights on all night can be placed into perspective. Feeling upset about wasting electricity is understandable but also doesn't make me a horrible human being.

CBT is often criticized for ignoring the role a person's past plays in their current struggles, but this is not entirely accurate. A CBT clinician would acknowledge the role a major life event such as a life-threatening disease played in shaping our present cognitive-behavioral repertoire (e.g. Persons, 2008). The way Beck approached this was through another concept he developed known as schemas (1967), which he referred to as the set of subtle beliefs, subtle in that we are largely unaware we hold them, about the world, ourselves, and the future, which set the mold for the kinds of automatic thoughts and cognitive distortions to which one is vulnerable.

Schemas develop early in life and are shaped by both experiences/ environmental factors and genetic predispositions (Persons, 2008). For example, a child who is genetically prone to anxiety and is raised by overprotective parents might come to have a schema that the world is a dangerous place and we can never be safe in it. The schemas that develop early in life can later be activated by the events that transpire. Returning to the example of the person who was surprisingly kind when I crashed into their car, Beck's theory would predict that they might have had a supportive environment as a child and genetic factors that promoted resilience that led them to see the world as a better place.

Schema activation is the way in which a life experience triggers or awakens our schemas into the action of spouting out interpretations or thoughts that in turn strengthen these beliefs (Persons, 2008). For instance, seeing photos of my friends hanging out without me on social media might activate a schema that I am unlovable or that others cannot be trusted, which initiates a cascading sequela of thoughts and behavioral reactions. These responses, while on the surface appearing to help us cope with the stressor, can ironically strengthen maladaptive schemas. If I were to react to my friends' photos by ignoring my friends out of spite or losing my cool with them, they might push me further away.

Cognitive behavior therapy aims to modify or restructure thoughts to render them more realistic but also incorporates relaxation strategies to modulate physiological manifestations of emotional responses as well as behavioral modifications such as teaching problem solving.

CBT and Eco-distress

Climate change has prompted a myriad of responses, influenced by a complex web of factors that are cultural, interpersonal, contextual, as well as individual in nature. Research shows that beliefs in the world being a dangerous place negatively affects mental health (Clifton & Meindl, 2021). In relation to a CBT case conceptualization (Persons, 2008), described in the Introduction of this text, this would be a schema, as well as a mechanism that would be directly targeted in treatment.

A belief in the world being a dangerous place is common among clients I've worked with, and it makes sense to target this belief in clients who are neglecting self-care out of empathic concern for the world's problems, including climate change. Marks and Hudson (2024) recommend assessing whether eco-anxiety is constructive as opposed to impairing, meaning it is associated with unconstructive worry and rumination.

The idea of constructively responding to eco-anxiety goes hand in hand with Victor Frankl's concept of *tragic optimism* (Stulberg, 2024), according to which you can experience suffering and also hold onto hope at the same time.

Tragic optimism (Stulberg, 2024) can be cultivated by validating the client's existential concerns while also working collaboratively to create at least a tiny sliver of hope in a better world and a role that they can comfortably play in creating a better world. Individual pro-environmental actions such as composting or avoiding food waste are examples of small ways to start. Sharing your concerns about global atrocities with others is another way, but it must be balanced with the recognition that the benefit of one person's concern is too small to risk their emotional or physical health, and therefore one might need to take a break from negative news.

How we see ourselves and the world also has a profound effect on our social relationships, which, as has been emphasized throughout this book, are pivotal to the success of collective efforts. Beck's approach can be used to help people target cognitive distortions, such as catastrophizing and personalizing when they can't, let's say get signatures on a petition, or don't fit in with an environmental group. This will help the person sustain their emotional resources and problem-solve (the behavioral component of CBT) to help them find other solutions.

Clinicians can validate their clients' fears about the effects of climate change and their frustration in response to the neglect of institutions toward the climate crisis. At the same time, clients can be led to understand when it is appropriate to exert energy toward trying to fix the issue (i.e. when refraining from engaging in fast fashion) and when they are wasting valuable time and resources (i.e. by ruminating). A negative world view can be modified by encouraging the person to balance doomscrolling with positive news about environmental solutions.

Schema Therapy

Schema Therapy is an intervention that is similar to Beck's approach but focuses exclusively on schemas (e.g. Arntz et al., 2021). Developed by Jeffrey Young, Schema Therapy entails finding common schemas that are associated with a "need state" that presumably developed as a result of negative early childhood experiences in which psychological needs weren't met. According to Young, compensatory or "conditional" schemas represent automatic coping behaviors to avoid schema activation, which leads to the unfortunate consequence of the need not being met. If I feel dismay over an environmental group's social media photos when they don't include photos of me being part of the group, I might avoid confronting other members about feeling excluded in order to prevent further exclusion (or further activation of my unlovable schema), which leads me only to feel more disconnected socially.

In Young's approach (Edwards, 2022), the therapist attempts to "attune" to the client's schema by having the client retell a recent event when difficult emotions were experienced, and presumably, schemas were activated, and helping the client sit with and weather through the challenging thoughts and feelings associated with the activated schema and unmet needs.

While not meant to invalidate concern about climate change, drawing attention to activated schemas can help people who are feeling stymied in pro-environmental efforts. Concern over whether peers in an advocacy group like them or think they are competent enough, for example, might represent an activated schema of never being good enough. This is something that has certainly affected my efforts at collective action. Every time I join a group, there is some criterion I don't seem to meet, like not having done enough research, clinical work, or been well versed enough in political or environmental science. When I work on understanding my "not good enough" schema and sitting with it, I can at least stick with the group or hope I will find another one that better meets my needs.

Why might some people have an easier time than others finding their place in a cause? We can look at models used by cognitive behavioral therapists to help explain why some adapt better than others. Here we focus on the following approaches from the biopsychosocial perspective as they pertain to eco-distress: The Diathesis Stress model; Differential Susceptibility Model; Social Learning; and Life Course models. In addition, Bronfenbrenner's Ecological Systems Theory and Persons' Cognitive Behavioral Conceptualization are also relevant to this discussion but are described in the Introduction to the text, so they are reviewed only briefly here.

DIATHESIS-STRESS MODEL

The diathesis-stress model has been framed as a way to understand the psychological effects of the tremendous stress experienced by climate scientists (Calabria & Marks, 2024). According to the diathesis-stress model, one's biological dispositions (the diathesis) blossom into a mental health disorder if the individual is sufficiently triggered by an environmental event (stress) such as trauma, maladaptive parenting practices, poverty, etc. The more severe the environmental stressor, the more likely people are to develop psychological symptoms even if their genetic profile predisposes them to emotional hardiness. On the other side of the coin, those with a greater predisposition to developing mental illness (i.e. a greater diathesis) might develop a mental illness in response to typical life stressors, such as academic demands, chores, and traffic jams (e.g. Nevid, 2022). The example mentioned above presumes that climate scientists are under stress severe enough to elicit distress even in those who are not predisposed to mental illness.

The diathesis stress model can work well as a way to conceptualize other examples of eco-distress. Some examples of situations to which this model would apply include individuals who are victims of traumatic natural disasters exacerbated by climate change (i.e. people who lose relatives in forest fires or severe floods), as well as those whose livelihoods are destroyed due to climate change, such as drought destroying a farmer's income. Individuals experiencing less severe climate change stressors (i.e. exposure to scientific knowledge about climate change dangers) might be prone to eco-distress if they are predisposed to being more emotionally sensitive.

DIFFERENTIAL SUSCEPTIBILITY MODEL

The differential susceptibility model (e.g. Assary et al., 2023) resembles the diathesis-stress model in emphasizing varying levels of predisposition to mental health problems. Where it differs is in highlighting that those who are more vulnerable, while at greater risk of disorder in the face of stress, also hold the greatest potential to benefit from prevention efforts. For example, incorporating social emotional learning into an elementary school curriculum will benefit those who are vulnerable to social-emotional problems more than it will benefit those who are not predisposed to these problems. We can apply this model to efforts many climate psychology professionals have made to creating educational materials for elementary school students to help them understand climate change in a way that helps them feel less scared and more empowered (e.g. https://mochi4theplanet.com).

BANDURA'S SOCIAL LEARNING THEORY

Bandura's social learning theory proposes that young children are actively processing social information (Rathus, 2022). Processing begins with observation, particularly with attention to relevant stimuli. For example, for a child to learn how to cope with frustration by observing their parents, they need to be paying attention to their behaviors. A key feature of social learning theory is that the child's modeling of behaviors is usually not automatic. As seen in Bandura's classic Bobo doll experiments. children mimicked aggressive behaviors they watched adults enact toward the dolls, but only when they themselves were presented with the dolls. Similarly, outside of the laboratory session, a child might not model their parents' behavior, such as aggressive behavior, until contextual aspects of their environment (e.g. being teased) and internal correlates (thoughts and emotions that make aggression more likely) align to open the door for the modeled behavior. This assumes, thus, that the child's observations are encoded in memory and remain latent until circumstances provide an outlet for them.

Social learning has both direct and indirect implications for pro-environmental behaviors. For one, we are more likely than not to inherit the values of our parents, and values correlate with behavior, such as a value of nature going hand in hand with conservation. Also, adults have great power to influence nature-friendly hobbies, like hiking in the woods, and to model resilience behaviors like problem solving over avoidance. Parents can validate their child's emotions and set limits in a collaborative way, such as asking a child to weigh in on how the family can prevent food waste or conserve water usage in the home (communication with Linda Goldman, May 28, 2025).

LIFE COURSE MODELS

An understanding of life course models, such as critical and sensitive developmental periods and the chain-risk or pathways model can enrich health professionals' case conceptualization of individual patients as well as inform community level prevention efforts (Wagner et al., 2024). Here, I discuss the concept of prenatal sensitive periods, during which a fetus' development is particularly sensitive to environmental agents, including toxins, encountered by the mother.

I experienced two pregnancies, one in 2013 to 2014 and the other in 2016 to 2017. While avoiding raw fish and cold cuts, and moderating caffeine intake were common knowledge, not a single health professional advised me to avoid air pollution as much as possible, even though research has strongly shown that it contains toxins that are carcinogenic (e.g. Brumberg et al., 2021). When we had a gas leak in our home and I shared my concern that it might have affected my pregnancy with my son, my doctor shrugged it off by saying anything could potentially affect an unborn child but there was no way of knowing whether my experience had.

This might in part be because it is impossible to completely avoid it. After all, you need to drive to work and might have to get older children on the school bus. This speaks to the need to address our culture of individualism alongside providing medical information to the public. For example, imagine a cultural shift in which parents advocate electric buses or simply carpool more.

If more intensive public education efforts shared the health ramifications of pollutants from school buses, parents would be banging at the doors of the superintendent's office at the blink of an eye to demand replacing fuel burning buses with electric ones. As long as it remains a distant, vague, environmental concern, we will lack the immense power that parents have to influence culture and political policy. CBT therapists can validate parents who are concerned about pollutants and encourage them to take action such as speaking up for their rights to a cleaner environment, even if in smaller ways such as offering to help a pregnant mother with tasks that might expose them to more toxins.

BRONFENBRENNER'S ECOLOGICAL SYSTEMS THEORY

The applicability of Bronfenbrenner's framework (Bronfenbrenner, 1979) to eco-distress is discussed in Crandon et al. (2024), where the authors conceptualize the microsystem as including attitudes and experiences of family and friends.

Crandon et al. (2024) conceptualize the mesosystem as including community resources and responses to climate change and shared communal stress; the exosystem as including messages from the media and policies instituted by governing bodies; and the macrosystem as encompassing cultural beliefs and the threat to cultural practices posed by climate change and environmental devastation. The chronosystem reflects living in the era in which the threat of climate change has become more present in the media and its effects more apparent.

These developmental aspects can be integrated into a CBT case formulation such as the framework offered by Persons (2008), described below, and can also complement other developmental theories such as social learning theory. For example, parental attitudes can encourage environmentalism (microsystem); parents and teachers can help their children access community resources, such as reading books from the library about environmentalism (mesosystem). In the exosystem, climate change and environmental science can be included in school curricula, and pro-environmental practices can be instituted in the workplace.

These practices in turn can gradually shift cultural attitudes and beliefs about environmentalism. My description has placed parents at the center stage, but the responsibility for making societal changes should not solely fall on their shoulders. Teachers and other school and community officials as well as religious institutions and other pillars of the community can play a role as well. In individual therapy, the practitioner can guide the client in accessing these resources as well as interacting with others successfully toward the effort of serving a role in the community. They can help the individual navigate conflicts that arise from others who don't hold the same values.

Mental health providers, including CBT therapists, can also provide auxiliary services to their clients and the community. In addition to his clinical and administrative duties, James McKowen, a clinical psychologist and co-chair of the Climate Change and Mental Health Initiative at Massachusetts General Hospital, leads climate cafes or informal gatherings that provide safe spaces for people to express their emotions related to the climate crisis (communication with J. McKowen, August 8, 2025). He also offers lectures to the public on topics such as the effects of extreme heat on mental and physical health. Through agencies that help disadvantaged youth, he talks with teenagers about climate emotions and the benefits of green space. These services are separate from his role as a cognitive-behavioral therapist working with teens and adults struggling with addiction, although he also provides psychoeducation about heat to his clients, particularly those who work outdoors and who struggle with substance abuse, which can negatively affect thermoregulation.

CBT Case Conceptualization

As outlined in the Introduction, Persons (2008) lays out the following elements of a CBT case conceptualization: Origins of mechanisms, which include genetic factors like temperament and early childhood experiences; mechanisms, including schemas as well as behaviors like avoidance and perfectionism; precipitants or events that activate maladaptive beliefs/schemas, and problems that result from this activation, like depression.

When it comes to actively engaging in climate work, extreme thinking and the fear of failure can easily become obstacles given we're faced with adversity from a society set up to encourage consumerism and pollution. This can underly a distressed reaction to a precipitating event such as learning about a natural disaster in the news.

All-For-Nothing Thinking and Eco-distress

In a *Psychology Today* post (Keluskar, 2024), I refer to the distress caused by All *for* nothing thinking. As opposed to all or nothing thinking, which is a cognitive distortion reflecting rigid thinking more broadly, all for nothing thinking is the fear that one will put great effort into something without getting the hoped for outcomes in return.

In the post (Keluskar, 2024), I refer to the stoic challenge or interpreting adversity as a challenge to practice coping in a way that is in line with your values. The idea of stoicism is particularly relevant to climate distress, although there are myths we hold about it.

Bill Irvine, author of *The Stoic Challenge: A Philosopher's Guide to Becoming Tougher, and More Resilient* (2019), confronts the common misconception that stoics want to be miserable. He contends that in contrast, stoics want to be comfortable and will alter their expectations in order to be more content. In an email exchange in which I asked Irvine how stoicism relates to perfectionism, Irvine responded that stoics do not tend to be perfectionists, but rather, they follow the principle of "good enough." Dr. Irvine explained, as an example: "Are you hungry? A meal in a fancy restaurant is nice, but bread and water is good enough" (personal communication, January 2022).

Accepting the possibility of not achieving hoped-for outcomes can be helpful in preventing climate advocacy burnout. The idea is that climate change requires urgent action, and at the same time the widespread societal changes we need to make will necessarily be slow moving. The important thing to hope for is that you will continue to have bursts of mental energy to keep plugging away at trying to make a difference to whatever extent possible. Along the lines of the behavioral part of CBT, an individual can be guided in budgeting their mental energy in order to prevent all-for-nothing thoughts that fuel exasperation and are more likely to hinder you when you are tired.

Automatic thoughts, cognitive distortions and schemas addressed by the CBT clinician, need not focus specifically on climate change to help clients engage proactively in this existential crisis. For instance, several of my clients have felt a loss of meaning in their work while simultaneously working long hours toward being successful in hopes that they would save enough income to afford donating to a non-profit or having the time to devote to volunteer efforts. Why the contingency? Could we as clinicians help clients create sustainable work-life balances toward the end of helping them carve out more time and energy for creative and humanitarian causes? I believe we can, but it will take more than telling them to join an environmental group or recycle.

The CBT model allows us to look at how our interpretations about ourselves, our relationships with others, our future prospects, and the nature of the world itself feed into thoughts, emotions, and behaviors that hold us back. For example, assertiveness skills might be helpful to the person who is too timid to ask for a non-disposable cup to drink out of. Tackling extreme thinking and promoting a more flexible mindset (Marks & Hudson, 2024) might help prevent burnout in climate activists. Helping someone with obsessive compulsive disorder overcome their contamination fears might allow them to live in line with their pro-environmental values by using fewer resources for excessive washing.

Addressing perfectionism, resentment, avoidance, extremism, and poor self-care allows people to make progress toward their environmental goals, feel part of a community, and stay active in their efforts over the long term.

My own perfectionism and anxiety prevented me from having healthy relationships with other environmental advocates, leading me to either give up passively or to lose my cool after increasing resentment spilled over. I've found that CBT strategies have helped with addressing these personal obstacles.

THE LIMITATIONS OF CBT FOR ECO-DISTRESS, AND ADJUNCTS THAT CAN HELP

In their comprehensive chapter on adapting CBT for eco-distress, Marks and Hudson (2024) acknowledge how CBT falls short of addressing climate emotions.

For one, they note conventional CBT's lack of cultural sensitivity and propose measures such as the Cultural Formulation Interview (APA, 2013).

Acknowledging one's cultural history and practices can intersect with approaches used by eco- and art therapists involved in the climate space, such as forest bathing, and using naturally degrading Earth Clay, leaves, and twigs as an artistic expression of feelings and an avenue for expressing difficult thoughts and feelings (communication with M. Keshet, October 2023 and July 28, 2025).

Mor Keshet, an Integrative Eco-Art Therapist and the Long Island, New York Regional Coordinator of the Climate Psychology Alliance of North America (CPA-NA) applies the framework of *biomimicry* with complex trauma clients by having them immerse themselves in nature and use natural materials to express themselves (communication with M. Keshet, July 28, 2025).

Emotional Biomimicry (Keshet, 2025) is the idea of learning about the order of the natural world to inspire greater resilience. Keshet describes how, rather than label emotions as "good" or "bad," it is more constructive to see all emotions as informative and to tap into the strengths of our nervous system for responding to them adaptively. She uses analogies rooted in nature to teach her clients emotional resilience, such as invoking the image of ducks with oil-coated, waterproofed feathers, which render them capable of shaking, ruffling, and diving in the face of storms.

Second, Marks and Hudson (2024) acknowledge the limitation of CBT's therapeutic emphasis on fixing maladaptive emotions can be inappropriate. Uncomfortable emotions about climate change should be validated, supported, and embraced. Our hope as a society is that eco-distress will become the norm, not a syndrome. On the same note is CBT's focus on correcting illogical thinking. There is a tricky balance the clinician must perform effectively between avoiding inappropriate minimalization of the climate crisis – how can you call scientific facts stating that we are at risk of extinction "catastrophizing"? – and addressing thoughts that unnecessarily fuel hopelessness and disengagement, such as, "What's the use of trying to change anything if we are doomed?"

Last, Marks and Hudson (2024) contend that the focus of therapy should shift away from personal internal beliefs and toward one's ecological beliefs and identity. Part of the work of a CBT therapist should also entail cultivating acceptance toward those who struggle to get along with their loved ones who are not feeling the same level of concern about the climate crisis.

Finally, existential angst needs to be acknowledged alongside suggestions for coping adaptatively and creating a hopeful outlook. This can go hand in hand with teaching the client to engage in meaning-focused coping (Ojala & Chen, 2024). For example, a clinician can help a client find meaning in the challenge of navigating tensions with friends and family who are not as concerned about the climate crisis, or frustration with other environmental advocates who don't seem as invested in taking action.

Psychotherapists must rise to the occasion in ramping up the psychological hardiness needed to be part of what is still considered, among many, to be an unpopular solution: Sustainable measures and dramatic cultural change.

CONCLUSION

When I first became interested in climate change psychology, I interacted and collaborated with other cognitive behavioral clinicians who heard the calling to use their professional skills towards creating a safer, more sustainable world.

At this time, there is no official treatment manual for eco-distress from a cognitive-behavioral perspective, and training is limited in using CBT to help people with this concern. However, momentum is building in this area. For example, clinical psychologist Thomas Doherty offers a training for mental health professionals (https://selfsustain.com/) in which he incorporates principles from Acceptance and Commitment Therapy (see Chapter 8) and concepts from eco-psychology. There are numerous workshops being offered that discuss perspectives that could be adjunctive to CBT, such as psychological first aid for victims of natural disasters.

I hope that I have given the reader more tools and insights, as well as ideas to inform future research on manualized treatments.

REFERENCES

American Psychiatric Association. (2013). *Diagnostic and statistical manual of mental disorders* (5th ed.).

Arntz, A., Rijkeboer, M., Chan, E., Fassbinder, E., Karaosmanoglu, A., Lee, C. W., & Panzeri, M. (2021). Towards a reformulated theory underlying schema therapy: Position paper of an international workgroup. *Cognitive Therapy and Research, 45*, 1007–1020. https://doi.org/10.1007/s10608-021-10209-5

Assary, E., Krebs, G., & Eley, T. (2023). Practitioner review: Differential susceptibility theory: Might it help in understanding and treating mental health problems in youth? *The Journal of Child Psychology and Psychiatry, 64*(8), 1104–1114. https://doi.org/10.1111/jcpp.13801

Beck, A. T. (1963). Thinking and depression. I. Idiosyncratic content and cognitive distortions. *Archives of General Psychiatry, 9*(4), 324–333.

Beck, A. T. (1967). *Depression: Clinical, experimental and theoretical aspects.* Harper and Row.

Beck Institute: Cognitive Behavior Therapy. (2025). Dr. Aaron T. Beck. *About Beck Institute.* https://beckinstitute.org/about/dr-aaron-t-beck/#earlylife

Bronfenbrenner, U. (1979). *The ecology of human development: Experiments by nature and design.* Harvard University Press.

Brumberg, H. L., Karr, C. J., Bole, A., Ahdoot, S., Balk, S. J., Bernstein, A. S., Byron, L. G., Landrigan, P. J., Marcus, S. M., Nerlinger, A. L., Pacheco, S. E., Woolf, A. D., Zajac, L., Baum, C. R., Campbell, C. C., Sample, J. A., Spanier, A. J., & Trasande, L. (2021). Ambient air pollution: Health hazards to children. From the American academy of pediatrics policy statement. *Pediatrics, 147*(6), e2021051484. https://doi.org/10.1542/peds.2021-051484

Calabria, L., & Marks, E. (2024). A scoping review of the impact of eco-distress and coping with distress on the mental health experiences of climate scientists. *Frontiers in Psychology, 15*, 1351428. https://doi.org/10.3389/fpsyg.2024.1351428

Clifton, J. D. W., & Meindl, P. (2021). Parents think – incorrectly – that teaching their children that the world is a bad place is likely best for them. *Journal of Positive Psychology, 17*(2), 182–197. https://doi.org/10.1080/17439760.2021.2016907

Crandon, T., Thomas, H., & Scott, J. (2024). Social-ecological perspectives and their influence on climate distress in young people. In K. Hudson & E. Haase (Eds.), *Climate change & youth mental health: Multidisciplinary perspectives* (pp. 287–308). Cambridge University Press.

Merriam-Webster. (n.d.). *Cope.* Retrieved April 11, 2026, from https://www.merriam-webster.com/dictionary/cope

Dozois, D. J. A., Dobson, K. S., & Rnic, K. (2019). Historical and philosophical bases of the cognitive-behavioral therapies. In K. S. Dobson & D. J. A. Dozois (Eds.), *Handbook of cognitive-behavioral therapies* (4th ed., pp. 3–31). Guilford Press.

Edwards, D. J. A. (2022). Using schema modes for case conceptualization in schema therapy: An applied clinical approach. *Frontiers in Psychology, 23*, 763670. https://doi.org/10.3389/fpsyg.2021.763670

Hofmann, S. G., Asnaani, A., Vonk, I. J. J., Sawyer, A. T., & Fant, A. (2012). The efficacy of cognitive behavioral therapy: A review of meta-analyses. *Cognitive Therapy Research, 36*, 427–440. https://doi.org/10.1007/s10608-012-9476-1

Irvine, W. B. (2019). *The stoic challenge: A philosopher's guide to becoming tougher, calmer and more resilient.* W.W. Norton & Company.

Keluskar, J. (2024, June 29). All-for-nothing angst. *Psychology Today.* https://www.psychologytoday.com/us/blog/alone-together/202406/all-for-nothing-angst

Keshet, M. (2025, September 12). *Post-traumatic ecology: Learning emotional resilience from the living world.* Biomimicry Institute. https://biomimicry.org/post-traumatic-ecology-learning-emotional-resilience-from-the-living-world/

Marks, E., & Hudson, K. (2024). Cognitive behavioral principles for conceptualizing young people's eco-emotions and eco-distress. In K. Hudson & E. Haase (Eds.), *Climate change & youth mental health* (pp. 148–167). Cambridge University Press.

Nevid, J. (2022). *Essentials of psychology: Concepts and applications* (6th ed.). Cengage.

Ojala, M., & Chen, X. (2024). Coping with climate change among young people: Meaning-focused coping and constructive hope. In K. Hudson & E. Haase (Eds.), *Climate change & youth mental health: Multidisciplinary perspectives* (pp. 269–286). Cambridge University Press.

Persons, J. (2008). *The case formulation approach to cognitive behavioral therapy.* Guilford.

Rathus, S. (2022). *Childhood and adolescence: Voyages in development* (7th ed.). Cengage.

Stulberg, B. (2024, November 29). How not to fall into despair. *The New York Times.* https://www.nytimes.com/2024/11/29/opinion/tragic-optimism.html

Wagner, C., Carmeli, C., Jackisch, J., Kivimäki, M., van der Linden, B. W. A., Cullati, S., & Chiolero, A. (2024). Life course epidemiology and public health. *Lancet Public Health, 9*(4), e261–e269. www.thelancet.com/public-health

Whalley, M. G. (2019). Unhelpful thinking styles: Cognitive distortions in CBT. *Psychology Tools.* Retrieved July 26, 2025, from https://www.psychologytools.com/articles/unhelpful-thinking-styles-cognitive-distortions-in-cbt/

Chapter 10

Conclusion

Final Thoughts for Helping People Find Their Role in Climate Solutions

In an episode of Julia Louis Dreyfus' podcast, *Wiser Than Me*, the then 84-year-old House Speaker Nancy Pelosi, who served as speaker for the House of Representatives from 2007 to 2011, shared tips for staying resilient in the workforce (Dreyfus, 2024). A few simple, yet powerful pieces of advice from Pelosi helped sustain me as I made several transitions in my professional and personal life over the past few years: (1) Know your power, (2) know your "Why," and (3) be ready.

Pelosi's (Dreyfus, 2024) point about remembering my why, though, is what has allowed me to stay committed to finding a way to do something about climate change even when I felt I didn't have anything to offer in the way of climate solutions. Before I sat down to write this book, I posted blogs on mental health, which helped me hone my writing skills. I continued providing cognitive behavioral therapy (CBT) and gained insights into working with existential concerns. All the while, I repeatedly failed at climate activism. But advocacy opportunities can come knocking on your door in many forms if you keep working on the projects you feel passionate about.

KNOW YOUR POWER

While writing this manuscript, I had the opportunity to speak with a climate advocate named Lauren Krueger, to whom I refer in Chapter 5. Lauren is part of Dayenu, a Jewish climate action group. I wanted to get more details about her story.

DOI: 10.4324/9781003489665-11

Lauren shared that she channeled her frustration with the political situation in the United States into a desire to give back to her community:

> I needed a project. I was a stay-at-home mom at the time ... I needed a way to connect to the world again ... I didn't know how I could contribute or where I could make, if not a difference, just some small amount of change somehow.

Lauren then found out about Field Team 6, a group of ex-reality television producers who wanted to get more people to register and vote as Democrats. She leaned into her skill set by doing video editing for them, including for Paula Poundstone's *YouTube* shorts (personal communication, July 22, 2025).

However, she yearned for a creative project focused on climate change. As part of the board of her local synagogue, she later saw an opportunity when a new Rabbi asked what the synagogue needed. Lauren suggested a sustainability committee. Their first project was creating a garden on the synagogue's grounds. This was a clever way to start given that it was relatively simple and could unite people who cared about environmental sustainability with those who also had a green thumb and/or wanted to make the synagogue's grounds more beautiful. At present, Lauren has a small core group of committee members.

While acknowledging that it can be challenging to find time and energy to take on additional projects, Lauren has empowered her group by collaborating with Dayenu, an example of the kind of coalition building that activist and scholar Leo Goldsmith recommends for empowering one's cause (see Chapter 3). With the support of a colleague at Dayenu, Lauren got the word out about the connection between heat waves and climate change in a piece featured in the *Long Island Herald* (Krueger, 2024). The article emphasizes how larger corporations have tried to blame the individual for their carbon footprint without taking accountability for their own exponentially larger damage.

Lauren's story is a great example of knowing your power. Lauren doesn't have a background in politics, law, or climate science, but she engaged in the causes she supported through her skill sets in video editing and writing. She then saw an opportunity to voice her concerns about the environment through her role in her synagogue. She gradually grew her cause by joining forces with Dayenu.

KNOW YOUR WHY

Knowing my "why" has been pivotal to the progress I have made in finding a way to contribute to environmentalism. Whenever a client brings up an existential concern such as climate change, it has reminded me of the importance of informing clinicians of the subtle but important ways in which the climate crisis can affect mental health. Remembering the importance of fighting for a cleaner earth to ensure a better future for my has allowed me to persist in finding my place in climate psychology as well. And when it came to persisting with this book? Well, I knew it was a risk given all we still need to learn about incorporating CBT into climate solutions. But we need to start somewhere.

Throughout my career, I've noticed a tendency to want to tread waters into which not many have ventured. I lamented over the observation that the people who come up with ideas first don't tend to be the ones who are most successful. Rather, it tends to be those who take things that have been started already and further develop the initial ideas. I can't seem to get rid of this aspect of my personality, but I've learned to accept it and reframe what I thought was lack of success to what is more accurately a different version of it. If I contribute to putting ideas out there that others can then further develop, my contribution will be good enough, and I'm grateful for that.

BE READY

Being ready means diving into your passions even when you don't know how the outcome will turn out.

More health professionals are expressing the need to address the growing angst revolving around climate change and environmental devastation, and at this point it is inevitable that we will need to adapt to warming temperatures and the myriad effects this will have on our lives in the years to come.

If this book helps others find their place in doing what is most healthy for them to cope with the climate crisis, whether this be meditative or proactive, that alone will allow me to feel like it was worth the years of labor I put into these pages.

A thorough cognitive behavioral therapist will begin their work with assessment, not only to confirm any mental health diagnosis but also to consider formative events, demographic factors, relationships, belief systems, cognitive, and personality factors, and other aspects of identity in understanding the individual in front of them.

In this book, I analyzed various aspects of identity and how these aspects might affect one's beliefs, emotions, and behaviors surrounding climate change.

Chapter 1

In Chapter 1, I conceptualized different ways people have responded to the climate crisis. I described several "types" for illustrative purposes, to enable the reader to appreciate the diversity of behavioral repertoires evoked by the climate change crisis. These types include activists, academics, preppers, burrowers, victims, and strivers. This is meant as a road map to understanding the meaning that the climate crisis might hold for the individual and determining how well these responses are working for them in terms of their wellness and effectiveness.

Chapter 2

Chapter 2 explored intergenerational differences and similarities, offering perspectives on the opportunities for synergy when people of different ages understand each other's perspectives and ways of approaching climate solutions. Characteristics that intersect with age and affect climate change attitudes and behaviors were also explored, including perfectionism and perception of time.

Chapter 3

Chapter 3 explored other aspects of identity, including political and ideological stance, socioeconomic status, gender identity and sexual orientation, personality, neurodivergence, culture, and racial/ethnic background. Research exploring how these aspects might affect environmental attitudes and behavior highlight trends that will be important for health practitioners to be aware of and incorporate into assessment and intervention.

Chapter 4

Chapter 4 explored different world beliefs that individuals tend to hold, such as optimistic and pessimistic ways of seeing the world, individualistic versus collectivistic cultural backgrounds, and other psychological concepts that can be used to help a client lean into hope and build self- and collective-efficacy in working toward climate solutions. Eco-friendly philosophies were also explored in this chapter, as were the obstacles posed by culture, such as consumerism, and their psychological correlates, such as emotional insecurity and fear of missing out or FOMO.

Chapter 5

Chapter 5 dove into how spirituality plays a role in climate change responses, including describing religion and spirituality, the pros of inspiring hope, community, and effective responses (i.e. collective advocacy), and the potential dangers of becoming too engrossed in unscientific dogma, as well as the unbridled power that can come from religious authority figures. At the same time, being open to learning from different religions can be a way of understanding how belief systems inspire a positive, reciprocal relationship with nature, with humans as the stewards of nature as opposed to its usurpers. This included a discussion about incorporating eastern philosophies into helping people effectively process climate emotions. I proposed ways to map these principles into third-wave CBT approaches (DBT, ACT, and RODBT) which draw upon eastern philosophy and spiritual practices.

Chapters 6 through 9 focused on different CBT approaches and how they can be applied to the climate crisis, including which behaviors might also be contributing to obstacles clients face in engaging in environmental work and how to help the individual find their place in climate solutions.

Chapter 6

Chapter 6 covered Radically Open Dialectical Behavior Therapy (Lynch, 2018), an evidence-based intervention for mitigating problems of overcontrol such as inflexible thinking patterns, maladaptive perfectionism, a tendency to ruminate, be resentful and jealous of others, and have trouble forming intimate relationships.

Chapter 7

Chapter 7 covered Dialectical Behavior Therapy (DBT; Linehan, 1993), describing how this approach can address the effect of collective invalidation from governing bodies who minimize the climate crisis (J. Homan, personal communication, November 4, 2023). The application of DBT skills, including acceptance, mindfulness, interpersonal, and emotion regulation skills to helping people effectively engage in collective action was discussed.

Chapter 8

Chapter 8 covered Acceptance and Commitment Therapy (ACT; Hayes & Strosahl, 2005), which focuses on being able to approach your thoughts and feelings as distinct from the essence of your being, trying to understand them with a spirit of curiosity, and seeing them as phenomena as opposed to definite fact. ACT also entails learning to live in accordance with your values or overcoming obstacles that are getting in the way of doing so. The discussion extended to research and writing about environmental identities and values, and techniques that have been adapted from ACT for helping activists avoid burnout.

Chapter 9

Chapter 9 covered conventional CBT. A CBT case conceptualization or formulation includes an understanding of biological predisposition to mental health difficulties, and how biological and early environmental factors contribute to the formation of schemas or beliefs we hold about ourselves and the world, which in turn frame how we interpret events that transpire. The limitations of CBT for eco-distress were also explored as well as how to overcome these limitations.

This book expanded CBT to understanding and addressing eco-distress, and in so doing I hope it gave precedence to the effect one's surroundings have, through societal, institutional, geographical and interpersonal factors, on our ability to find meaning and purpose in our chosen cause, specifically the climate crisis.

Let's work together on working together. Who knows what we could achieve if we could really understand each other, appreciate our differences, and figure out how to create a better world together. We have nothing to lose by dreaming big and helping each other with the challenges (and opportunities) that lie ahead.

REFERENCES

Dreyfus, J. L. (2024, December 18). Julia gets wise with Nancy Pelosi. (Season 3, Episode 9 [Audio podcast episode]. In *Wiser than me.* Lemonada.

Hayes, S. C., & Strosahl, K. D. (Eds.). (2005). *A practical guide to acceptance and commitment therapy.* Springer Science + Business Media. https://psycnet.apa.org/record/2005-02327-000

Krueger, L. (2024, August 9). To beat the heat on Long Island, we must fight fossil fuels. *LI Herald.com.* https://www.liherald.com/stories/heat-climate-change-fossil-fuels-krueger,209350

Linehan, M. (1993). *Cognitive-behavioral treatment of borderline personality disorder.* Guilford Press.

Lynch (2018). *Radically open dialectical behavior therapy: Theory and practice for treating disorders of overcontrol.* Context Press.

Index

For Product Safety Concerns and Information please contact our EU representative GPSR@taylorandfrancis.com Taylor & Francis Verlag GmbH, Kaufingerstraße 24, 80331 München, Germany

Batch number: 10476187

Printed by Printforce, the Netherlands